THE FISHERMAN'S BUSINESS GUIDE

THE FISHERMAN'S BUSINESS GUIDE

by Frederick J. Smith

International Marine Publishing Company
Camden, Maine

TO ANNE

CONTENTS

PREFACE

Some years ago, fishermen came to Oregon State University seeking assistance in the management of their fishing businesses. This initial interest led to the development and presentation of fishing business management extension courses to several thousand fishermen from Alaska to Florida and from Maine to California. Somewhere along the way, it became obvious that most of the fishing business management course material could be shared more widely and efficiently if it were available in a permanent printed form.

This book is the result. Some suggestions on its use are noted in the introduction. I should point out that this book includes only information and ideas that are unlikely to become obsolete in a few years' time. There is no discussion of tax regulations, resource management regulations, or ongoing research. This type of information has a high rate of obsolescence and is readily found in numerous softcover publications, many of which are listed in the bibliography.

An excellent working environment was provided by the Agricultural Economics Department and the Marine Advisory Program at Oregon State University. A similar environment also existed in the Resource Economics Department and Marine Advisory Program at the University of Rhode Island while I was on leave at that institution. A number of colleagues have contributed in various ways to the compilation of this work. Drs. Harvey Meier, A. Gene Nelson, James Fitch, and Frank Conklin at Oregon State University provided advice on early drafts of the first eleven chapters. Dr. Virgil Norton of the University of Rhode Island contributed to the accuracy of Chapter 1, and Professor Andreas Holmsen, also of the University of Rhode Island, contributed to the content and applicability of many chapters. Credit is also due Mrs. Dana Cramer, who patiently and efficiently typed the final manuscript. The assistance of each is gratefully acknowledged. As is customary, the author assumes responsibility for the content of this work.

Frederick J. Smith

INTRODUCTION

What makes a successful fisherman? What skills, talents, and knowledge must a fisherman have to be successful? Is it sufficient to be a good net man? Is mechanical ability important? Will skill in navigation lead to success in fishing?

All these are important, but we all know of good net men, good mechanics, and good navigators who have failed as fishermen. *Success in fishing, if we measure success in terms of profit, is determined by the fisherman's managerial ability.* Your ability to harvest great quantities of fish does not guarantee success. You must also make the right business decisions.

The primary objective of this book, therefore, is to help you become a good manager. The requisites for good management are:

1. An understanding of the economic, political, and physical environment within which you conduct your business.
2. An understanding of business-management principles, concepts, and tools.
3. The ability to use these principles, concepts, and tools in managing your fishing business.

How the Book is Organized

The organization of this book is based upon the requisites for good management. The early chapters discuss principles. Later chapters apply these principles.

Part I contains seven chapters. Two of these chapters deal with the economic, political, and physical environment within which you operate your fishing business. Overfishing, fishery resource management, fish prices, and marketing are all subjects treated widely in industry journals and discussed among fishermen. In spite of the apparent importance of these subjects, they deserve only two chapters in this book because they can be dealt with effectively only through group or collective action. Success in the fishing business can be accomplished through individual action — that is, good business management — in spite of overfishing, fishery resource management, fish prices, and marketing. The other five chapters in Part I deal with economic principles and concepts, knowledge of which will enhance your application of the management tools illustrated in Part II.

There are five chapters in Part II. Each chapter treats one of the important tools in successful fishing business management. These are: management information, profit and efficiency analysis, partial budgeting, financial budgeting, and business organization.

Part III illustrates business-management applications to some of the major fishing business decisions: selecting a fishery, boat size and ownership, and boat management.

How to Use the Book

First, this book *should* be used — each week, month, and year. It is not meant to be read and left on the bookshelf. However, a thorough reading, once through, will greatly enhance its value. Future reference can then be made to the chapter or section that will help you at that time.

The busy fisherman in search of some quick guidance on decision-making procedures can refer to Chapter 10. The student attempting to understand the decision-making process can study Chapters 2, 3, and 8. A state fishery biologist attempting to understand the behavior of individual fishermen can read Chapters 1, 4, and 5.

Although there is something in this book for nearly everybody, the book is directed primarily toward the nearly 100,000 commercial fishermen in the United States and Canada who own or operate one or more boats and are concerned with financial success. Any intelligent fisherman who is motivated to learn and succeed can master the material presented. It is not complicated, but neither is it simple. You will find unfamiliar words and ideas. You may have to reread some chapters to understand them better.

The reward for a better understanding of fishing business management can be significant — your own success.

PART ONE

ECONOMIC CONCEPTS

1 FISHERY RESOURCE MANAGEMENT

A common-property resource is the property of no one person, but of all people in common. The air is common property as are the waters of the oceans, and no person can legally deny another access to these resources. The living resources of the ocean are also viewed as common property. However, when these living resources are harvested, they have market value and are thereafter treated as private property.

There is a long history of interpersonal, interstate, and international conflicts regarding the exploitation of living ocean resources. Various levels of government have found it necessary to change some of the common-property characteristics of living ocean resources in order to resolve not only historical conflicts but also an increasing number of new ones.

As a fisherman, you depend upon this common-property resource for your livelihood and must operate your business within this changing and frequently unpredictable environment. Your ability to manage a successful fishing business will be enhanced by an understanding of the common-property nature of the fisheries and the attendant conflicts that must be resolved.

The Expansion of a Fishery

In an industry where all resources (such as land and labor) are privately owned, new firms will enter and overall production will be expanded as long as each firm is earning some profit.[1] As more privately owned resources are used, the cost of these resources tends to rise; and as more goods are produced, the prices of these goods tend

---

[1] All businesses use some common-property resource; for instance, air is used in operating an internal-combustion engine. However, a discussion of conflicts in the use of other common-property resources is beyond the scope of this book.

Production

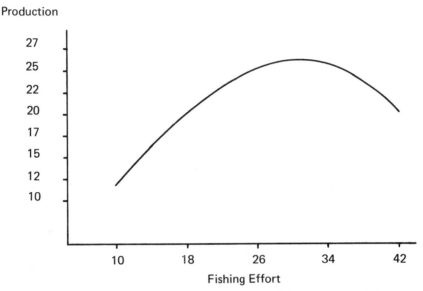

Fishing Effort

FIGURE 1-1.   Graph of the relationship between fishing effort and the annually sustainable production for a hypothetical fishery.

to fall. Therefore, profits for all firms in that industry must eventually decline as the industry expands. Nevertheless, the industry will expand as long as the return to the most limiting privately owned resource is at least as great in this industry as it would be if the resource were used in some other industry. Put simply, the resource is used where it will earn the most. When each firm earns a "reasonable profit," industry expansion is stabilized.

The most dominant characteristics of the fishing industry are very high or very low profits and instability. Yet the economic forces described above are at work in the fishing industry as in any other industry. The distinguishing difference is that the primary resource in the fishing industry is not private property — it is common property. How this difference affects you, the individual fisherman, can best be explained by using the concept of the fishery production function.

The Fishery Production Function

A production function shows the relationship between some output and some input. In a fishery production function, the total harvest of fish in a given season is measured as output; this harvest re-

TABLE 1-1.   Relationship between fishing effort and the annually sustainable
production for a hypothetical fishery.

Units of Fishing Effort	Annually Sustainable Production
10	120
18	200
26	250
34	260
42	200

sults from various levels of fishing effort, the input. Table 1-1 illustrates a hypothetical fishery production function — the various levels
of annually sustainable fish production resulting from various levels
of effort.[2]

In Table 1-1 we can see that as fishing effort is increased, total
sustainable production increases rapidly until it reaches 250 tons. At
this point, an additional 8 units of effort increases sustainable production to 260 tons. Another 8 units of effort actually decreases
production to 200 tons. This is made clearer in Figure 1-1, a curve
based on the data in Table 1-1.

The increase in production that accompanies increased effort is
the result of greater exploitation of previously unexploited stocks.
Your harvest of the fish is a substitute for natural mortality. At
some level of effort (34 units in our example), fishing reduces the
stock more than would have occurred as a result of natural mortality. That is, increased harvest begins to reduce parent stock so that
recruitment or replacement of harvested fish is at successively lower
levels. In our example, 42 units of effort harvests so much of the
fish stock that there are not enough parents left to rebuild that stock
from season to season. At 34 units of effort, there is exactly enough
parent stock left to maintain a production of 260 tons season after
season. This is referred to as the maximum sustainable yield.

---

[2]   Effort can be some index of the number of boats, number of people, units
of gear, and days of fishing in the particular fishery. Sustainable production
implies that the same effort will produce the same volume of fish next season
and every season thereafter.

The Problem

With an understanding of the fishery production function and the economic characteristics of an expanding industry, we can begin to understand the problem in fisheries. The problem is due to the common-property nature of the resource and is manifested as follows:

1. A total industry fishing effort beyond the maximum sustainable yield is the result of rational decisions by many individual fishermen, but collectively imposes a significant cost upon society.
2. The effort of one fisherman imposes an unanticipated or unaccounted-for cost upon all other fishermen, which influences their decisions regarding expansion.

The first point is perplexing, but politically intriguing. If you can see that profits are being made in a new shrimp fishery, it is logical for you to enter that fishery in quest of those profits. Your increased effort plus the effort of other fishermen may produce a total effort in excess of the maximum sustainable yield, yet still

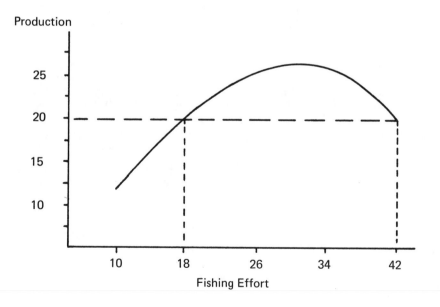

FIGURE 1-2.   Graph of the relationship between fishing effort and the annually sustainable production, showing the same production at two levels of effort.

generate a profit for each fisherman. Figure 1-2 makes this clearer. A total effort of 42 units produces 200 tons of shrimp; but notice that 18 units of effort will also produce 200 tons of shrimp. From society's viewpoint, does it make sense to use 24 additional units of effort to place the same amount of shrimp in the marketplace? The answer is no if the resources going into that additional effort can be better utilized elsewhere. Yet, there is nothing inherent in the fishery that would cause the accumulated effort of fishermen to be 18 units, or any other number of units for that matter. When added together, the effort of each fisherman who acts independently and rationally in quest of greater profits may total 10, 18, 26, 34, 42, or any other number of units. That is, there is nothing inherent in the fishery production function that precludes a profit at any level of production. If, however, the fishery resource were privately owned, new entrants in the fishery would be forced to bid up the price of the resource in order to obtain a share. This would increase greatly the cost of fishing, and profits to all fishermen would be gained only at lower levels of effort.

Without knowing exactly what the fishery production function looks like or what the total effort will be, various public agencies respond to this problem by "managing" the fisheries. Management can be directed at supplementing fish stocks — supported by general tax revenues or special levies from the fishing industry — or it can be directed at limiting fishing effort.

To date, limiting fishing effort has been accomplished by limiting harvest efficiency. The techniques are familiar to all fishermen: gear restrictions, closed seasons, closed fishing areas, fish-size restrictions, and so on. This approach to fishery management sustains most of the characteristics of a common-property resource, but conflicts directly with fishing business management (the management of an individual fishing business). While you strive for greater efficiency and production, our public agencies attempt to reduce your production by restricting your efficiency. You may find it difficult to abide by such management techniques, even with the realization that our public agencies are attempting to manage for the benefit of society as a whole.

With increasing domestic and foreign competition for existing fishery resources in recent years, more attention has been given to limiting fishing effort by limiting the entry of capital, labor, or both, into the industry. Although this may not conflict with the objectives of an individual fishing business, it compromises one of the primary characteristics of a common-property resource: no person can deny another access to a common-property resource. Some economists as well as fishermen feel that we have reached a point in the world-wide expansion of fisheries where the cost of managing the fisheries on a common-property basis is far greater than the benefits derived. If fisheries are managed as a private resource at some point in the future, the profitability of individual fishing enterprises is likely to be much more comparable to those of other industries. Before we reach this point, we must settle the very difficult question of who gains ownership of the various fishery resources.

Let us now consider the second way in which a common-property fishery affects the operation of a fishing business: the effort of one fisherman imposes an unaccounted-for cost upon other fishermen. Let's illustrate by assuming a fishery with 10 vessels, each of which consistently harvests 100 tons and makes a profit. Desiring a profit from the same fishery, we remodel our vessel and get in on the action. Depending upon the level at which this fish stock has been exploited, two things can happen. Our catch and the catch of the other 10 vessels will decrease to less than 100 tons, or we all must fish harder (spend more money) to maintain the 100-ton production levels. We all suffer because we all are taking fish from the same common-property resource. If we had recognized this fact and if we could have estimated the effects beforehand, we may not have entered this fishery and the profitability for the 10 fishermen presently exploiting this fish stock would have remained at higher levels. Also, if the fish were a private good and not common property, we would have included the cost of acquiring this good in determining the profitability of the fishery. Our estimate of profitability would then have been more accurate, and the decision to enter or not enter this new fishery would have been more realistic. The resulting number of fishermen, or the effort in the fishery, would also have been less.

Because the basic resource being exploited is common property and because individual fishermen fail to anticipate all the costs imposed by an additional participant in a fishery, overfishing is a constant threat to the industry. This threat and the measures imposed to counter this threat play an important role in fishing business management.

2 THE DECISION-MAKING PROCESS

Fishing success depends upon your skill in handling vessels, handling gear, finding productive fishing areas, and harvesting the fish you find. Financial success depends upon your decision-making ability as well as your fishing skill. Decision making is the most important part of business management. It can make the difference between prosperity and bankruptcy in fishing.

Decision-Making Ability

A story of two fishermen will help illustrate this important concept. One, whom we shall name Fleet Mind, began his fishing career as a crewman on one of the better boats in port. Swift Hand, the other fisherman in our story, was born into a prosperous fishing family and inherited his shrimp trawler debt-free. Because of radical differences in their decision-making abilities and therefore their management, Swift Hand concluded his fishing career as an able crewman on Fleet Mind's boat.

Swift Hand was serious and hard working. He was an excellent mechanic and could repair all types of gear. He understood boat design and construction. He spent more time in his boat's engine room than at the local pub or at home. How could Swift fail and Fleet succeed? In short, Fleet viewed his fishing career as a business venture, and based all his decisions on the financial pros and cons, whereas Swift considered fishing an occupation, where hard work and good, efficient equipment formed the basis of most of his decisions.

Fleet began building equity in a humble but seaworthy boat early in his career, whereas Swift immediately went into debt to buy a new, sleeker, and more specialized boat. Fleet remained flexible, switching from fishery to fishery as fish stocks and catch rates rose and fell and as market prices went up or down; but Swift responded to falling catch rates and market prices by investing in more exotic specialized gear and working longer, harder hours. Fleet looked for various ways to reduce costs while Swift was investing more money and time to get more power from his engine and more speed from his boat. Swift's deteriorating financial condition was not evident to the casual observer of the well-maintained boat and its hard-working skipper. Swift's inability to meet his debts eventually forced his creditors to foreclose, and Swift found himself a willing and capable employee of Fleet.

The experience of these two fishermen doesn't suggest that you will go bankrupt if you inherit your boat, nor does it suggest that you shouldn't work hard and maintain an efficient fishing operation. It does suggest that the financial success of your fishing business depends upon your ability to make decisions on a dollars-and-cents basis as well as on a technical or aesthetic basis. Fleet's decisions were more often the correct decisions, whereas Swift was often wrong. Fleet understood the decision-making process, but Swift failed to grasp a fundamental point: making decisions should make profits.

Five Steps in Decision Making

Decision making is a logical process. This process can be broken down into five steps, which illustrate the importance of decision making in the management scheme. Your experience may bring to mind additional illustrations and may suggest some different decision-making steps. A thorough understanding of the decision-making process will help you apply the economic concepts and tools discussed in following chapters. The five decision-making steps follow:

1. Make Observations and Obtain Ideas

You read an article in the *National Fisherman* regarding a new midwater trawl. The National Marine Fisheries Service (NMFS) publishes some data showing large stocks of midwater fish offshore. You go to look at the trawl, and you call the NMFS for more information on the new fish stocks. You are making observations and obtaining ideas.

2. Analyze Your Observations

You now use paper and pencil. You calculate the additional gear, horsepower, boat alterations, and so forth, that you require in order to use the new trawl. Using various tools illustrated in the following chapters, you calculate the increased operating cost, the loss in income from fisheries you will have to give up, and the anticipated increase in total receipts. You weigh the accuracy of your information and determine whether you should invest at this point in additional and more accurate information. You consider the increase or decrease in risk and uncertainty for your business, and weigh this against your financial condition.

3. Make the Decision

You have all the ingredients for an accurate decision: profit or loss, change in risk and uncertainty, and reliability of your decision-making information. You measure all this against your objectives — more profit, less risk, more leisure time, or whatever — and make the decision: yes or no.

4. Take Action

You arrange financing with the bank, place your order for the net, and busy yourself adapting the boat to the new midwater trawl; or, you put the information in the file and return to your previous fishing.

5. Accept Responsibility

If you have decided to buy and use the new net and if all the data provided you was reasonably accurate, you must accept the responsibility for the financial failure as well as the financial success of the project.

Gathering and Using Information

Fruitful ideas come readily to the manager with a practical imagination, an observing eye, and an inquiring mind. He watches the innovations of his fellow fishermen, reads the popular and technical fishing literature, and relentlessly pursues the advice and ideas of knowledgeable people around him.

It is rarely possible or even practical to obtain enough information to ascertain the anticipated results of a decision. Moreover, complete analysis is seldom practical. Nevertheless, it is important that the analysis be as accurate as possible, given the decision-making information available. The fisherman with the best information in the world will still go bankrupt if he doesn't know how to use that information. The successful manager has a complete set of management tools, and knows where and how to use them.

Making the decision is the simplest yet most difficult part of the decision-making process: there is almost always some risk and uncertainty involved, and the responsibility rests upon the decision maker. Some people cannot bring themselves to make a decision. Decisions in the fishing business take more courage than many possess.

Taking Action and Accepting Responsibility

A decision is not made until it is implemented. Most of us know a fisherman who has "decided" to quit fishing, yet we see him out fishing year after year. Not until the boat is sold and the fisherman is busy driving a truck has this decision been made. This may seem a minor point, but it illustrates the difference between the daydreamer and the man of action. The man of action follows the decision process to the end: he implements his decision.

Finally, there are fishermen who can't credit themselves enough if the decision is profitable and can't find enough scapegoats if the decision turns sour. Inability to accept responsibility for decisions marks the candidate for failure. He is likely to be the same fisherman who had great difficulty making the decision in the first place. Some people are better suited to working for others and leaving most of the decisions to the "boss." You should recognize which role suits you best.

Decisions Classified

There is an important difference between a decision regarding the purchase of a new boat and a decision concerning a return to port during a storm. Deciding upon the crew's share and the grade of oil for the engine require different amounts of your attention and concern. Likewise, you will react differently to an increase in ice cost than you will to a drop in fish prices. It is useful for you to distinguish among such decisions and to recognize which deserve more or less attention from you, the decision maker.

Decisions can be conveniently classified in terms of: (1) importance; (2) frequency; (3) imminence; and (4) revocability. Each is discussed below.

Importance

The importance of a decision depends upon the size of the loss or gain involved. Buying a new boat involves thousands of dollars and, to a large extent, the future of your business, but you are unlikely to go bankrupt if you return to port with a half load during a storm. Also, deciding upon the grade of oil for the engine will have much less financial impact upon your fishing business than a decision regarding the crew's share. You should allocate your scarce decision-making time according to the ultimate financial impact of each decision.

Frequency

You make some decisions so frequently that they become almost automatic. For example, your use of electronic navigational aids in setting a course can become so automatic that your decisions are subconscious. But if there is a hint of malfunction in your electronics and you are approaching a breakwater in a fog, these decisions suddenly become conscious and extremely important, no matter how frequently they are made.

Frequent decisions should become automatic or should be delegated to others in the crew, unless there is a change in the circumstances surrounding those decisions. However, such decisions may still be important because of their cumulative effect. Less frequent decisions can receive more of your attention if and when they arise, regardless of their importance.

Imminence

An imminent decision is one that must be made immediately — it can't wait. When your vessel is shipping water in a heavy sea, the decision regarding dumping some fish and heading for shelter had better be made quickly. It is a very important decision — hopefully not too frequent, but imminent when it arises. A less imminent decision may concern the next engine overhaul, the purchase of some new crab pots, or the welding of some extra flanges on your net drum. If you delay your decision for a few days, or even a few weeks, you are not likely to go bankrupt. The penalty for waiting before making a decision is high when the vessel is foundering but low if you are considering an engine overhaul. When the penalty for waiting is low, the time can be used to obtain more information and to analyze the decision further. Procrastination in this case can help increase the accuracy of the decision, but the decision maker must also be able to estimate the loss of profits due to delaying the decision.

Revocability

Once your pots are set, your nets cast, or your long line anchored, you have made a decision that can be revoked only at considerable loss. An even less revocable decision is to borrow $18,000 and install a new engine and refrigeration system in your vessel. A decision regarding the stowage of a purse seine is easily revoked, as is a decision regarding the type of lure on trolling lines. Obviously, decisions that can be revoked only at high cost deserve more attention in the decision-making process, but easily revokable decisions may be experimented with and may provide valuable information for future decisions.

3 BUSINESS OBJECTIVES AND PLANNING

Objectives, a plan for attaining those objectives, and the management ability to carry out the plan are prerequisites to a successful business, just as the identification of a good fishing area, appropriate navigational aids, and a reliable engine and rudder are prerequisites to successful fishing. In this chapter procedures for identifying business objectives are discussed, and the planning process is outlined. Detailed planning, in the form of budgeting, is discussed in Chapter 11.

Identifying Objectives

Most of us have some vague notion of what we want out of life. We may envision money, fame, happiness, or some combination of these. The objectives of your fishing business depend largely upon these personal objectives and the objectives of your family. The first step in identifying your business objectives is to identify your personal and family objectives. This can be a pleasant family affair in which all sit around the kitchen table and quiz one another regarding what each "wants out of life." It is your responsibility to integrate these diverse objectives and formulate consistent objectives for your fishing business.

This is not a once-in-a-lifetime or even an annual event. It is a continuing process. Not only do the objectives of your family change, but the environment within which you live and conduct your fishing business constantly changes. As you gain experience in your business and in life, your personal and business objectives are likely to change in relative importance. A younger fisherman with little equity and no family will find security a less important objective than short-run profit. A senior fisherman with much equity in his business will likely forego a quick-money, high-risk venture.

Possible Business Objectives

A fishing business may have a variety of objectives, among which may be the following:
1. Maximum profit.
2. Business survival.
3. A combination of profit and leisure.
4. Financial security.
5. Continued business growth.

Money is a common denominator in nearly all of these business objectives. This common denominator makes economic concepts and tools valuable to you, whatever your business objectives might be. Let's look at the five business objectives listed above in more detail.

Profit Maximization

Profit maximization means that the business is organized in all respects to produce the greatest profit possible. It means that the boat is fitted only with the gear or facilities that contribute to profit. It means that you sell your fish where you receive the highest possible price, and it means that you stay at sea as long as it is profitable to do so. Profit maximization is a less popular objective than many believe. How many fishermen will forego a few conveniences and luxuries in the wheelhouse when having a new boat built? How many fishermen will move from buyer to buyer after he has become acquainted with a particular buyer or has joined a fishermen's cooperative? How many fishermen can resist heading for home a bit early when they have been out for weeks and the weather is a bit rough? Few are willing to make the total sacrifice required for profit maximization. This objective is usually tempered by other objectives.

Business Survival

Business survival can be an important objective if you are operating with low equity and very little available cash. Opportunities that may yield high profits or result in possible losses are rejected, for they may not be consistent with survival. Financial security is a similar objective in that you avoid fishing activities that could produce a loss; but in this case you are usually operating from a more secure financial base, so that a loss would not necessarily result in bankruptcy.

Profit and Leisure

A combination of profit and leisure is probably a more common objective among fishermen than the objective of profit maximization. You want to make a good living, but you also want to be independent, you want to have time with your family and time to go hunting, and you may also want the newest or largest boat in port. "Good living" is often referred to as a threshold income. You behave like a true profit maximizer until you reach the threshold income, at which point several other objectives become more important to you.

Growth

Although profit and security are closely related to the growth objective, growth usually means that you want your business to become larger — larger boats, more boats, larger catches. This can be incompatible with the profit objective and even the security objective. More and larger boats may require more debt. Investing in your own business may not be the most profitable use of your capital, and larger debts increase the danger of bankruptcy from bad seasons.

Planning

Planning is projecting the future course of your business. Once you have identified your objectives, you must develop plans to guide your business and your decisions toward attaining these objectives. As objectives change, so must plans change — the process is simultaneous. Unfortunately, the pressure of daily decisions makes it difficult to do very much long-run planning. We must set aside a specific time for this, just as we set aside time for identifying objectives.

The Planning Horizon

Some objectives can be satisfied with very little time and effort, whereas others require planning over the major part of your lifetime. For example, a "day off" tomorrow is a short-run and readily attainable objective; but a "comfortable retirement" must be planned during most of your working career, and for most of us this requires some short-run sacrifices and some long-run anxiety.

In order to have a day off tomorrow, we must be sure the boat is secured, all the fish are unloaded, and the crew is notified. This doesn't require much planning or thinking ahead. The planning horizon in this example is only one day. To have a comfortable retirement, we should spend some time with the family to determine what can be given up in the short run. We have to consider various long-term investments that will pay during retirement. Do we invest in our own business, in the stock market, in retirement insurance? What about inflation? What kind of living standard will we desire at retirement? This objective requires considerable planning and thinking ahead, and the process is continuous. Long-run plans are more difficult because they involve more unknowns and require constant updating.

Some Steps in Planning

Broadly speaking, the accomplishment of any objective depends upon your available resources, your capabilities, and the economic, political, and social world within which you operate. Therefore, the first step in planning is to identify these factors clearly.

Available resources include those presently under your control, such as cash on hand, your gear, your boat, your crew, and your other investments. They also include resources that you can readily obtain, such as untapped credit, good relationships with processors or other fishermen, and a good reputation. Personal capabilities include fishing skill, management skill, technical skill, and the ability to acquire more of each. The economic, political, and social world includes such things as the present and future demand for seafood, the supply of fish in the ocean and estuary, inflation, seafood imports, written and unwritten laws and regulations, and social pressures.

The second step in planning includes a determination of the resources needed and the personal capabilities required to accomplish the objective. You must also consider the legality, social accept-ability, and economic feasibility of the plan. If your resources are inadequate, you must consider where, how, and if you can obtain additional resources. If you are not personally capable of carrying out your plan, you should determine where you can obtain help or how you can improve your skills. Finally, you must devise ways to satisfy legal constraints and you must adjust your plan to coincide with the economic realities of the world.

Step three in planning involves the application of many of the management tools discussed in the following chapters.

4 DEMAND, SUPPLY, AND FISH PRICES

There is probably more confusion over fish prices than over any other aspect of the fishing industry. Fish prices appear to go up and down in spite of imports, landings, and the best efforts of fishermen. In general, all prices are determined by demand and supply. However, in the fishing industry there are many factors that interfere with the normal price-determining nature of demand and supply. Several of these factors, as well as the concept of demand and supply, are discussed in this chapter. Basic information is provided for estimating prices, and therefore profitability, in different fisheries. This information may also help you understand some of the price statistics published by public and private agencies.

The Demand for Fish

When a consumer orders shrimp creole or finnan haddie in the restaurant, he is expressing his demand for fish. The restaurant operator (or supermarket manager) buys his supply of fish from a wholesaler or local distributor, who in turn buys from the coastal processing or fish plant, who in turn must buy from you, the fisherman (unless it is imported). Consumers create demand because they like the taste of fish, because it is nutritious, because it is cheaper than steak, and so forth, and this demand keeps you in business.

Suppose for the moment that you can buy a halibut steak dinner in the restaurant for $4.85. What happens when the price of a halibut steak dinner jumps for some reason to $5.35? Isn't it likely that you will eat less halibut steak? Other halibut eaters will react in the same manner. This results in an overall decrease in the amount of halibut demanded.

The opposite would be true if the price dropped to $3.95. You

TABLE 4-1.   Relationship between price of halibut steak dinners and number of dinners bought (hypothetical data).

Price of Halibut Steak Dinners	Number of Dinners Bought in May (Quantity Demanded)
$4.85	600
4.95	500
5.05	400
5.15	300
5.25	200
5.35	100

would probably buy halibut steak dinners more frequently, and others would join you. This results in an increase in the amount of halibut demanded. The relationship between these prices and quantities is called a demand schedule or demand curve. These are illustrated in Table 4-1 and Figure 4-1. The hypothetical demand schedule or demand curve simply illustrates more vividly what we already know: when the price of halibut dinners goes up we buy less, and when the price of halibut dinners goes down we buy more.

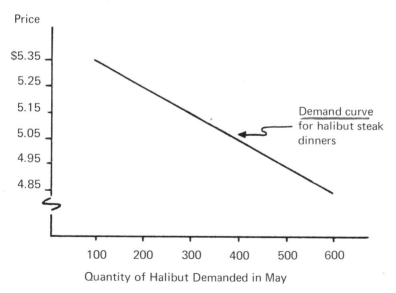

FIGURE 4-1.   Graph showing the relationship between the price of halibut steak dinners and number of dinners bought (hypothetical data).

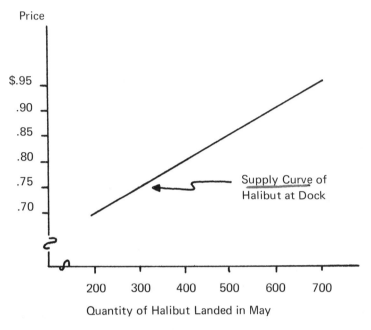

FIGURE 4-2.   Graph showing the relationship between price paid to fishermen for halibut and quantity of halibut landed (hypothetical
data).

So far, we have carefully avoided any comment regarding the
supply of halibut. In order to understand how the price of halibut
steak dinners became $4.85 in the first place, we must look at sup-
ply as well as demand.

The Supply of Fish

As a halibut fisherman you are likely to put more effort into
catching halibut if the dockside price is $.95 per pound than if it is
$.70. Also, as a New England flounder fisherman you are likely to
spend more days at sea and invest in more gear if landed prices
average $.32 per pound rather than $.12. Higher prices spur you to
greater effort, help you justify the higher costs, and, in the end, increase the quantity of halibut, cod, flounder, scup, scallops, shrimp,
oysters, salmon, and so on, being landed. Higher prices also attract
more fishermen into that particular fishery, and if the fish are available, landings are again increased.

As with demand, the quantity of halibut landed at different prices can be illustrated as a supply schedule or supply curve. Table 4-2 illustrates a hypothetical supply schedule and shows us that more halibut will be landed at higher prices and less halibut at lower prices. Figure 4-2 illustrates a hypothetical supply curve based on the information in Table 4-2.

The Marketing Margin

The difference between dockside prices and prices in retail stores or restaurants is referred to as the marketing margin. The marketing margin covers transportation costs, cost of risk, and profits of all the people who handle the fish between the docks and the consumer. In our halibut example, the difference between landed price and restaurant price increases from $4.15 to $4.40 per pound as dockside and retail prices increase. This pays for steaking, packaging, freezing, cold storage, trucking, preparation at the restaurant, other restaurant services included with the halibut steak dinner, and the profit of all who are involved in this marketing process.

We now have all the pieces of our price-determination puzzle.

How Demand and Supply Determine Fish Prices Under Ideal Conditions

Once you understand the demand and supply curves for fish, it is relatively easy to understand how prices are determined under ideal conditions. However, because these conditions rarely exist in

TABLE 4-2.   Relationship between price paid to fishermen for halibut and quantity of halibut landed (hypothetical data).

Price paid to Fishermen for Halibut	Quantity of Halibut Landed in May
$.70	200
.75	300
.80	400
.85	500
.90	600
.95	700

TABLE 4-3.   Relationship among retail price, fisherman's price, and quantity of halibut (hypothetical data).

Price of Halibut Steak Dinner		Marketing Margin		Price at Fisherman's Level	Quantity Demanded in May
$4.85	–	$4.15	=	$.70	600
4.95	–	4.20	=	.75	500
5.05	–	4.25	=	.80	400
5.15	–	4.30	=	.85	300
5.25	–	4.35	=	.90	200
5.35	–	4.40	=	.95	100

the fishing industry, the following discussion may appear theoretical. Nevertheless, demand and supply are always the strongest forces in determining price.

Using our hypothetical demand-and-supply schedule for halibut, let's see how the price would be determined at the fisherman's level. In Table 4-3, we subtract the marketing margin from the restaurant price to obtain the demand schedule at the fisherman's level. Using the halibut price at the fisherman's level, we can now show the quantity demanded and the quantity supplied for each price.

Notice in Table 4-3 that when the price is $.80, the quantity

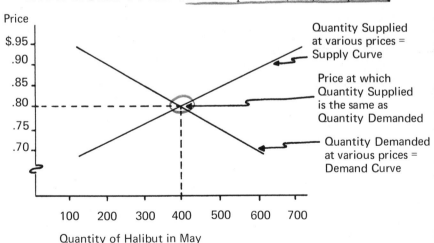

FIGURE 4-3.   Graph of the relationship among fisherman's price and quantities of halibut supplied and demanded (hypothetical data).

demanded is exactly the same as the quantity supplied. When the
what they want at the $5.05 restaurant price ($.80 plus $4.25). The
cold storage and halibut steak dinner eaters are able to buy exactly
what they want at the $5.05 restaurant price ($.80 plus $4.25. The
amount of halibut being landed is exactly the same as the amount
being bought.

Figure 4-3 is based on the information in Table 4-3. It illustrates
a demand-and-supply curve. For the moment, suppose the price were
$.85. Halibut fishermen would be motivated to spend more time on
the grounds, go farther, work harder, and in general spend more
money to land more halibut. Our supply curve tells us that they
would soon be landing 500 units of halibut instead of 400.

But at $.85, halibut consumers are less inclined to buy halibut
and will start eating more cod, hamburger, or crab. Our demand
curve tells us that they will then consume only 300 units of halibut
instead of 400. Figure 4-4 illustrates the situation when the price is
$.85. Fishermen are landing 500 units of halibut in May, but con-
sumers are taking only 300 units at $.85. That leaves us with 200
extra units aging in warehouses or on a halibut schooner! Obviously,
this situation will not exist very long because halibut processors and
wholesalers will drop the price to fishermen and attempt to sell the

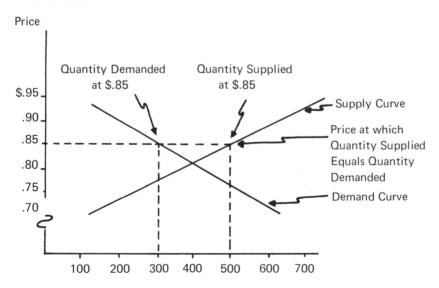

FIGURE 4-4.   Graph of the relationship between quantity of halibut supplied
                and demanded at $.85 (hypothetical data).

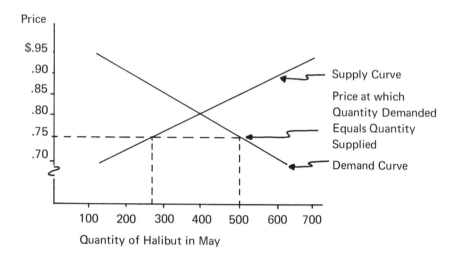

FIGURE 4-5.   Graph of the relationship between quantity of halibut supplied and demanded at $.75 (hypothetical data).

surplus at reduced prices. As processors and wholesalers cut prices, fishermen become discouraged and fish less, but consumers start buying more. This will continue until the price is again $.80. You can trace this pattern on the demand and supply curves in Figure 4-4; the demand and supply curves illustrate halibut consumers' behavior and halibut fishermen's behavior.

Now suppose the price were $.75. How would the demand-and-supply situation look? It is clear that consumers would be quite happy and would eat more halibut, but fishermen would reduce their halibut fishing and would probably look for other ways to make a living. Therefore, the quantity demanded would be greater than the quantity supplied. Figure 4-5 shows us that fishermen would supply 275 units and consumers would demand 500. There would be a shortage of 225 units. Obviously, this situation would not last for long because consumers, and therefore wholesalers and processors, would start competing with one another for the short halibut supply. One way to compete is to offer higher prices to fishermen. As prices are bid up, new fishermen enter the fishery, previous halibut fishermen fish harder, and more halibut is landed. Higher dock-side prices force up retail prices, and consumers reduce halibut consumption. This bidding up of prices, decrease in quantity demanded, and increase in quantity supplied will continue until the price is again $.80 and consumers are taking exactly what fishermen are landing.

Price Determination in the Sea-Food Marketplace

Sea-food demand and supply determine price. We know this
from our economic theory, which also tells us what the demand-and-
supply curve should look like. However, there are two important
factors operating in the sea-food market that make it difficult to
estimate demand and supply, and therefore prices. These two factors
also make prices appear to behave contrary to our expectations.

Dynamic Market

The sea-food market changes constantly. Consumers' tastes and
preferences for some sea foods change rapidly. This causes the entire demand curve for one sea food to shift left or right, causing
prices to fall or jump. (Shifting demand is illustrated in "Promoting
Consumption," a later section of this chapter.)

There are rapid and often unpredictable shifts in the supply of
sea foods. This is due to rapid changes in the available stock (caused
by a red tide, foreign fishing, hurricanes, a fish disease, and so forth).
It is also due to changes in fishing rules and regulations.

Changes in fishing rules and regulations frequently impose new
costs on fishermen. This would normally reduce supplies of fish, as
the fishermen with higher costs drop out of the regulated fishery.
However, we know from Chapter 1 that fishermen frequently under-
estimate the costs of fishing in a highly exploited fishery. Also, most
fishermen are persistently optimistic about making a profit during
the next year, in spite of higher costs. This results in a smaller than
normal decrease in supply.

In the sea-food market, the demand and supply curves do funny
things!

Market Structure

If there is to be a predictable and stable price in the sea-food
market, fishermen must have some accurate knowledge of consumer's desires, and retailers must have some accurate knowledge of
supply. Because of the historical structure of sea-food marketing,
processors, wholesalers, brokers, distributors, and retailers who have
not shared this knowledge with others have often been rewarded
with short-term profit. Therefore, the signals from consumers don't
always get through to the fishermen, and the signals from the fisher-

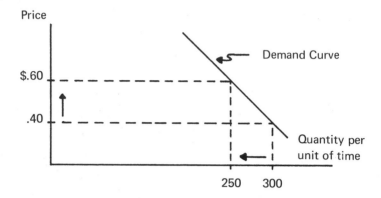

FIGURE 4-6.   Graph illustrating a 50-unit decrease in quantity demanded
with a $.20 increase in price (hypothetical data).

men don't always get through to the consumer. From time to time,
the signals that do get through may be wrong.

There are millions of sea-food consumers. There are over 100,000
commercial fishermen, but there are only several hundred sea-food
dealers. Some ports have as few as one dealer. In situations such as
this, there is little incentive for the dealer to increase the price to
fishermen when he can meet an increase in demand with his existing
supply and a greatly inflated price to the consumer. It is also possible for the fish dealer to absorb an increased supply by increasing inventories, greatly reducing prices to fishermen, and maintaining prices
to consumers.

Because of the market power of some sea-food wholesalers, geographic factors, tradition, and lack of interest by fishermen, there
has been profit in misinformation — overstated inventories, overstat-

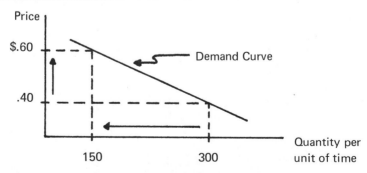

FIGURE 4-7.   Graph illustrating a 150-unit decrease in quantity demanded
with a $.20 increase in price (hypothetical data).

ed landings, misinterpreted import data, and so forth. This makes price predictions difficult and price stability unlikely.

Fish Prices and Profit

Fish prices are one of three important elements that determine your profit:

Profit = [Fish Price × Landings] - [Costs]

Although it would be nice to increase both fish prices and landings, our understanding of demand and supply indicates that this may not always be possible. An increase in price may mean that we cannot sell all we harvest. Nevertheless, we may still be able to increase profits. Fish price times landings gives us our gross stock or gross returns. If the percentage of increase in price is greater than the percentage of decrease in landings, our gross returns will increase.

Let's go back to the demand curve to illustrate what happens. Suppose our fishermen's association was able to increase the fish price from $.40 to $.60 per pound, as shown in Figure 4-6. According to the principles illustrated earlier in this chapter, we would expect a decrease in the quantity demanded. If the decrease in quantity is from 300 to 250, as shown in Figure 4-6, we wind up with a larger gross return, as shown below:

Price		Quantity		Gross Returns
$.40	X	300 units	=	$120.00 before
$.60	X	250 units	=	$150.00 after

A recent study by the National Marine Fisheries Service[1] indicates that this would be the case in several of our American fisheries. Although consumers will buy less lobster at a higher price, the price increase offsets the consumption decrease.

This same study also indicates that the opposite would be true for other fisheries. For example, suppose that our fishermen's association were able to increase the fish price from $.40 to $.60 and that the quantity demanded declined from 300 to 150 units, as illustrated in Figure 4-7. The price increase would not offset the

[1] The Division of Economic Research of the National Marine Fisheries Service has estimated price-quantity relationships for most American sea foods. These studies are listed in the bibliography.

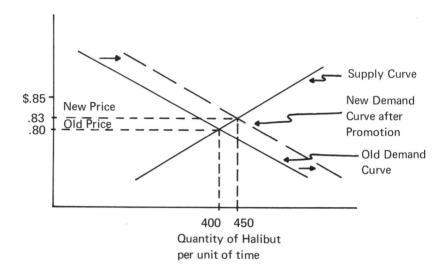

FIGURE 4-8.   Graph illustrating an increase in demand, new quantity demanded, and new price for halibut (hypothetical data).

consumption decrease, and gross returns would decline, as shown
below:

Price		Quantity		Gross Returns
$.40	×	300 units	=	$120.00 before
$.60	×	150 units	=	$ 90.00 after

Unless costs drop significantly, we are in worse shape with a higher
price than we were with the lower price.

What can a fishermen's price-bargaining association do? Most
important, it should learn what a price increase will do to members'
gross returns. If it is clear that gross returns will be reduced, you
might consider increasing the overall demand for your pollack, mullet, salmon, crab, and so forth. This can be done by promotion.

Promoting Consumption

Earlier, we described the demand curve as a picture of the quantities of fish that consumers will buy at various prices. If consumers
can be induced to buy more fish at any price, we will have changed

their demand; that is, the demand curve will have been moved to the right. It also means that consumers will pay more for any quantity of fish demanded. We can illustrate this by looking at the halibut demand and supply curves in Figure 4-8.

We have convinced consumers that halibut is more desirable. They will pay more for the same quantity, or will buy more at the old price. This is illustrated by the new demand curve in Figure 4-8. The new price, which will make the new quantity demanded equal to the quantity supplied, will be $.83, and the new quantity will be 450 units. This will benefit the fishermen's association if other halibut fishermen don't step in and exhaust this new market and if promotion costs are less than gross returns. Finally, we must remember that the processors and wholesalers may reap the benefits of this promotion for themselves and not allow price increases to be passed on to the association members.

5 FISHING COSTS AND RETURNS

"Costs," "expenses," "deductions," "payments" are words of small concern when you are writing your weekly or monthly checks. "Returns," "revenue," "earnings," "income" are words of small concern when fish are piling up on deck or in the hold. However, when preparing financial statements, analyzing profits, budgeting, and making business decisions, your understanding of these terms and their use is important.

In this chapter, we will explain each of the various costs and returns terms and then demonstrate four ways in which costs and returns are allocated in a fishing business.

Cost Classifications and Definitions

Fishermen, boat owners, accountants, bankers, and economists have their own cost terminology. Understanding these different terms can be time consuming and frustrating. Nevertheless, the fisherman who cannot distinguish between variable costs and fixed costs, cash costs and noncash costs, average cost and marginal cost, and so forth, will be handicapped in managing his fishing business.

The following is a partial list of common "cost" terms with brief definitions. You will see many of these terms in later chapters.

Costs

This is a very general term that applies to any or all of the terms below. The outlay of money or other things of value in exchange for goods or services received is a cost. The outlay of money may be only a "paper" or bookkeeping transaction, or it may be a promise to pay at a later date.

Expenses

Probably the only difference between a cost and an expense is in the person using the term. "Expenses" is more of an accounting term whereas "costs" is more of a management term. Expenses are less likely to include "paper" transactions.

Variable Costs

Costs that increase or decrease with fishing effort or production are variable costs. In a business where the relationship between costs and production is highly predictable, this term refers to those costs that vary with production. In the fishing business, where the cost/production relationship is less predictable than the cost/effort relationship, "variable costs" usually refers to costs that vary with effort. This distinction is illustrated later in this chapter and in Chapter 6. "Operating costs" is a term used interchangeably with "variable costs."

Fixed Costs

Any cost not included in the variable cost category is a fixed cost. Fixed costs do not increase or decrease with fishing effort or production; they are associated with ownership of the business and its assets. Fixed costs are sometimes referred to as "overhead costs" or "ownership costs."

Total Costs

Variable costs plus fixed costs equal total costs. However, total costs may include some of the costs described below that are not included in variable or fixed costs.

Trip Costs

This is a common fishing term that applies to those costs incurred during each fishing trip. All trip costs are variable costs, but many variable costs are not trip costs. For example, food, bait, and fuel are trip costs and variable costs, but engine repairs and crew-share are variable costs and not trip costs.

Boat Costs

Any cost incurred because of boat ownership is a boat cost. Boat costs may be variable costs or fixed costs. Repair and maintenance is primarily a variable cost because it varies with effort, but it is also a boat cost because it is associated with ownership. Depreciation and insurance are fixed costs because they don't vary with effort, and are boat costs because they are associated with ownership.

Opportunity Costs

This is an economic concept that you will find very useful. Put simply, an opportunity cost of one activity is the income given up by not participating in an alternate activity. Suppose you have $1,000 to invest and you identify two alternate investments: an FDIC savings account that pays a straight 6 percent annual interest, or an uninsured bond that pays 8 percent and costs $10 per year to guarantee payment. The opportunity cost of the bond is the $80 you could have earned from the FDIC savings account. Further examples of opportunity costs will be provided in this and later chapters.

Cash Costs

This is an accounting term that refers to payments actually made by cash, or some other thing of value.

Non-cash Costs

This is another accounting term; it refers to costs charged to the business "on paper" but for which payments are not actually made. The most common non-cash cost is depreciation. It is recorded as a cost in the "books," but no check is written to pay for depreciation.

Average Costs

Economists developed this term to measure efficiency. It applies to any of the above costs divided by the units of any input to the fishing business or any of the products. Average variable costs may apply to variable costs per ton of fish or to variable costs per day fished. Average fixed costs may apply to fixed costs per dollar invested or to fixed costs per unit of gear fished.

Marginal Costs

This is another very useful economic concept. It refers to the increase or decrease in total costs resulting from an additional day's fishing effort, or an additional ton of fish landed, or an additional crewman, or an additional unit of gear, and so forth. When matched with a resulting increase or decrease in gross returns, marginal costs show the economist (and the fishing business manager) the most profitable level of effort, amount of fish in tons, number of crewmen, units of gear, and so forth.

Understanding Variable and Fixed Costs

Variable costs in a fishing business usually include crew-share and fuel, ice, bait, food, and unloading costs. Fixed costs usually include insurance, depreciation, moorage, legal fees, accounting fees, association dues, and interest payments on business debts.

The repayment of principle on a business loan is not considered a variable or fixed cost. It represents the return of something borrowed and must be taken into consideration when making management decisions, but it is not a cost. The interest paid is the cost — it is the rent for using someone else's capital.

We have ignored one of the largest cost items — repair and maintenance. It deserves special treatment because some repair and maintenance can be treated as a variable cost and some as a fixed cost. Some repairs are due to natural deterioration, whether or not the boat is fished. While sitting at the dock, metal rusts and corrodes, paint flakes, planks rot and swell, and tides splinter wood. Costs incurred in repairing the damage by rust and corrosion to paint, planks, and wood are fixed costs because they have no relation to fishing effort or production.

However, the more you fish, the more frequently you will have to repair and replace shafts, winches, hatch covers, radios, propellers, bearings, blocks, cables, nets, and so forth. These are variable costs because they depend upon fishing effort.

Typical variable and fixed costs for a North Atlantic dragger are illustrated below:

Variable Costs

Boat repair and maintenance (70% of total)	$	6,200
Fishing and electronic gear repair (90% of total)		4,800
Fuel, oil, and filters		8,100
Food		3,500
Ice		3,200
Unloading		1,900
Crew-share		64,000
Total Variable Costs	$	91,700

Fixed Costs

Interest on investment	$	8,000
Depreciation		6,700
Insurance		5,200
Boat repair and maintenance (30% of total)		2,700
Fishing and electronic gear repair (10% of total)		530
Legal and accounting		500
Moorage		450
Total Fixed Costs	$	24,080

Total Costs	$115,780

In this illustration, some of the boat repair and maintenance costs are classified as variable costs (70 percent) and some as fixed costs (30 percent). Also, 90 percent of fishing and electronic gear repair is classified as variable and 10 percent as fixed. Such an allocation is somewhat arbitrary, but is important because it can have an effect upon your fishing decisions.

The importance of classifying costs as fixed or variable will be demonstrated in later chapters, but a simple example using a typical North Atlantic dragger will be helpful at this point. Assume that this is your boat and you are about to enter the 1976 season. You know that you will have costs of at least $24,080 (total fixed costs), whether or not you fish during 1976. You gather some information on fish stocks, market prices, and fishing competition, and you estimate that you could produce $130,000 worth of fish in 1976. Will you fish? It will "pay" you to fish because $130,000 is greater than your projected total costs of $91,700 plus $24,080, or $115,780.

Suppose, however, that you estimate a $100,000 gross return. Will you fish? If you do fish, you will cover all estimated variable costs and some of the fixed costs and sustain a loss of $15,780. If you don't fish, you will have a loss of $24,080, the total fixed costs. You are better off fishing. Your losses are reduced by $8,300 ($24,080 less $15,780) if you fish.

As long as you can cover all variable costs, it will pay you to fish. Something will be left over to help cover the fixed costs. This is not to say you shouldn't be concerned with the fixed costs, but as long as you own the boat and continue in the business, you will incur fixed costs regardless of the fishing effort. At some point in the future, you will have to decide about boat ownership and/or continuing in the fishing business.

The principle illustrated is an important one and deserves repeating: *As long as gross returns are greater than variable costs, losses will be minimized by fishing*. We will return to this when we discuss partial budgeting and boat management (Chapters 10 and 15).

The Opportunity Cost Concept

Opportunity cost was defined earlier as income foregone because you participated in one activity rather than an alternate activity. Some examples will help clarify this concept and illustrate its use.

The Opportunity Cost of your Time

You can use your time productively in several ways. You could work in a factory, you could work for the government, you could work in a boatyard, or you could go fishing. If you choose to use your time fishing, you can determine the opportunity cost of your time in fishing by determining the net income you would have realized if you had taken that factory, government, or boatyard job. Suppose that the government job would have given you $14,000 in take-home pay and that this is more than you could have earned in the factory or boatyard. By fishing, you are giving up the $14,000 take-home pay you would have earned had you worked for the government. This is a $14,000 cost because it represents income foregone or income you could have but did not receive. It is the opportunity cost of your time in the fishing business.

The Opportunity Cost of Investment

You can use your capital productively in several ways. You could invest it in the stock market, you could invest it in a savings account, or you could invest it in real estate, a hardware store, or a fishing boat. Suppose you have $50,000 burning a hole in your pocket and you decide you want to invest it in a boat. The opportunity cost of such an investment is the amount you could have earned in the *most profitable* and *equally risky* alternative. Although not the most profitable alternative, the hardware store probably represents a similar risk. The $8,000-per-year profit you could have obtained from owning the hardware store (not working in it) is the opportunity cost of investing in a fishing boat.

One of the easiest ways to arrive at the opportunity cost of your fishing business investment is to answer this question: "If I sold my business, boat and all, to an equally competent fisherman and held the mortgage myself, what is the minimum acceptable interest I would take?" If you would do it for 12 percent but no less, 12 percent is the opportunity cost of your investment.

Returns Classifications and Definitions

There are nearly as many different terms that refer to returns as there are cost terms. Some of the more common terms used by fishermen, boat owners, economists, and bankers are listed and defined below.

Returns

This is a very general term that applies to the receipt of money or other things of value resulting from the provision of service or delivery of goods to others. "Returns" may apply to any of the following terms.

Gross Returns

Gross returns is the total amount of money or things of value received during the specified time period (usually one year). In a fishing business, gross returns are calculated by multiplying price by total volume of fish landed. "Gross receipts" is synonymous with "gross returns." "Gross receipts" is more frequently used by accountants; "gross returns" is more frequently used by economists. "Gross revenue" and "gross stock" are other synonyms of "gross returns" and "gross receipts."

Gross Income

This includes all gross returns and gross receipts, but may also include money received from non-fishing business sources and from the sale of capital assets. It applies to personal matters as well as to fishing business management.

Gross Earnings

This is very similar to gross income, but usually applies to the money received from one's labor rather than from the productivity of an investment. "Gross earnings" is more applicable to the non-boat owner.

Gross Pay

"Gross pay" is similar to "gross income" and "gross earnings" but is used almost exclusively for money received from one's labor or management, before any deductions are made.

Net Stock

Gross stock less trip costs yields net stock. Net stock is used to determine wages (crew-share) for employees on the boat.

Net Returns

This applies to gross returns less any one or more of the costs defined earlier. Its meaning depends upon its use. "Net revenue" is the same as "net returns."

Net Income

Gross income less any one or more of the costs defined earlier yields net income. As with "net returns," its meaning depends upon what costs are subtracted.

Net Earnings

This is earnings less any one or more of the costs defined earlier.

Profit

"Profit" is a widely used term with as many meanings as users. However, it is generally understood that profit is the positive result of subtracting some costs from gross returns over a given time period.

Gross Profit

Gross receipts less all or some operating costs yields gross profit. This is an accounting term used most frequently in wholesale or retail businesses in which the greatest cost is in acquiring the goods to be sold. Gross profit in such a business is gross receipts less the cost of the goods to be sold.

Net Profit

This is another accounting term; it is determined by subtracting operating and overhead costs from gross receipts. Net profit is gross profit less the rest of the costs not subtracted above.

Return to Labor, Management, and Equity

Gross returns less all costs except those associated with the operator's labor, management, or equity is the return to labor, management, and equity.

Return to Labor, Management, and Investment

Gross returns less all costs except any interest on debts and costs associated with the operator's labor and management is the return to labor, management, and investment.

Average Returns

Gross returns divided by the units of one or more fishing inputs (days fishing, dollars invested, units of gear, and so forth) yields average returns. Gross returns divided by production yields average price. "Average revenue" means the same as "average returns."

Marginal Returns

A very useful economic concept, "marginal returns" is the increase or decrease in gross returns resulting from an additional day's fishing, an additional ton of fish landed, an additional crewman, an additional unit of gear, and so forth. "Marginal revenue" means the same as "marginal returns."

Applying Costs and Returns

Your use of costs and returns terminology will depend upon your immediate objective. You may wish to:
1. Calculate crew-shares, skipper-share, and boat-share at the end of a trip or season.
2. Calculate cash available to settle debts and sustain your family.
3. Calculate net profit, as defined by the IRS (Internal Revenue Service) for reporting federal income taxes.
4. Calculate annual business profits.

Each of these calculations is different and results in different numbers, even though they all apply to the same business.

Suppose you are the owner and skipper of a 72-foot North Pacific halibut schooner and have just completed the halibut season. You have maintained good records, a good ship's log, and a good diary. You now wish to use this information for each of the above objectives.

Example 1.   Calculation of crew-shares, skipper-share, and boat-share for a hypothetical 72-foot halibut schooner.

Gross stock	$125,800
Less: Lost gear	2,010
	$123,790
Less: Boat-share of 24% × $123,790	29,709
	$ 94,081
Less: Condemned gear	540
Ice and bait	12,000
Fuel	3,700
Food	3,300
Net crew-share	$ 74,541
Boat-share (from above)	$ 29,709
Less: Skipper's share of 10% × $29,709	2,971
Owner's share	$ 26,738

If you are the skipper and owner and have a crew of five, each crewman (including yourself) has a gross pay of $74,541 divided by 6, or $12,423. You have the gross earnings of your own crew-share ($12,423) plus the boat-share ($29,709), for a total of $42,132. If you are the skipper and not the owner, you have gross earnings of $12,423 and the skipper's share of $2,971, for a total of $15,394; the owner has a gross income of $26,738.

Example 2. Calculation of cash available to settle debts and sustain your family.

Gross returns		$125,800
Less:	Crew-share for five	62,115
		$ 63,685
Less:	Boat repairs	6,100
	Gear repairs	1,150
	Gear replacement	2,550
	Ice and bait	12,550
	Fuel	3,800
	Food	3,110
	Insurance premium	1,300
	Social Security contributions (crew)	2,700
	Miscellaneous	1,100
		$ 29,325
Less:	Interest on business debt	2,000
	Principal on business debt	4,000
		$ 23,325
Less:	Interest on personal debt	600
	Principal on personal debt	2,000
	Family living	14,275
Cash available		$ 6,450

After paying all cash costs incurred in your fishing business, $29,325 remains. After you pay the interest on business debts and meet your obligation for the return of principal, you have $23,325 in cash. After paying principal and interest on personal debts and after sustaining your family, you have $6,450 in cash to invest or enjoy. This is your true cash situation, and as you can readily see, the numbers are different from those derived in calculating crew-share, skipper-share, and boat-share.

Example 3.   Calculating net profit for tax reporting.

Gross receipts or sales	$125,800
Less: Cost of goods sold or operations:	
labor	62,115
materials and supplies	21,460
other costs	1,100
Gross Profit	$ 41,125
Other income (rental of gear)	9,000
Total Income	$ 50,125
Less: Depreciation	4,500
Repairs	7,250
Insurance	1,300
Interest on business debt	2,000
Net Profit	$ 35,075

The above example follows IRS Form 1040, Schedule C: "Profit (or Loss) From Business or Profession." The $35,075 net profit figure is entered on Form 1040 and added to other income in arriving at adjusted gross income.

In example 4 (page 44) we have not included the $2,000 in actual interest paid under fixed costs, and have therefore calculated return to labor, management, and investment rather than return to labor, management, and equity. The return to labor, management, and equity is $25,375 less $2,000 in actual interest paid, or $23,375. Return to investment ($13,375) less $2,000 likewise yields an $11,375 return to equity. Any of these measures is a good indicator of business profits. These will be dealt with in some detail in Chapter 9.

Each of the above examples results in different numbers; each has different objectives and uses different calculations. In Example 1, there is a boat-share of $29,709 and an owner's share of $26,738. In Example 2, there is $23,325 in cash after business cash costs have been paid. In Example 3, there is a net profit of $35,075 to report for federal income tax purposes. And in Example 4 there is a $25,375 return to labor, management, and investment.

Different costs and returns calculations and different results may cause confusion. However, confusion can be minimized if you: (1) understand the terminology used, (2) have the objectives of your calculations well in mind, (3) understand the calculation procedures, and (4) recognize the usefulness of different results.

Example 4: Calculation of annual business profits.

Gross returns	$125,800
Less: Variable costs:	
boat repairs	6,100
gear repairs	1,150
gear replacement	2,550
ice and bait	12,000
fuel	3,800
food	3,110
crew-share for crew of five	62,115
Social Security contribution	2,700
	$ 32,275
Less: Fixed costs:	
hull, protection, and indemnity insurance	1,300
depreciation	4,500
miscellaneous	1,100
Return to labor, management, and investment	$ 25,375
Less: Opportunity cost of investment	
($40,000 at 9%)	3,600
Return to labor and management	$ 21,775
Return to labor, management, and investment	$ 25,375
Less: Opportunity cost of labor and management	12,000
Return to investment	$ 13,375

6 MAXIMIZING PROFIT IN A FISHING BUSINESS

In Chapter 3, we discussed several fishing business management objectives. Profit maximization was one of the important objectives. Economists have recognized the importance of profit maximization and have developed basic economic principles that can be used in attaining this objective. In this chapter, we will explain and illustrate several profit-maximizing economic principles.

These principles may seem abstract and difficult to apply at first. However, a general understanding of them will increase the usefulness of the management tools illustrated in later chapters. They will guide you in developing management information, analyzing your profit, budgeting, analyzing your finances, and making decisions.

Production, Costs, Returns, and Profit

Costs, returns, and profit were defined in Chapter 5. The fishing business is dynamic: costs, returns, and profit change with changes in production, effort, gear, boat size, and other management factors. Moreover, costs may change more or less than returns, and profit may increase or decrease with changes in production, effort, gear, and so forth.

We will first discuss the relationships between production and costs, production and returns, and production and profit. We will then look at the relationships among other factors such as effort, gear, and boat size, and costs, returns, and profit.

Production and Costs

Next season's production (volume of fish landed) will depend upon the following:

TABLE 6-1.   Relationship between eight clam production levels and total
costs, given the season, gear, and boat size (hypothetical data).

Bushels of Clams Produced	Total Costs
0	$ 4,000
2,000	14,000
4,000	21,000
6,000	26,000
8,000	30,000
10,000	33,000
12,000	42,000
14,000	63,000

- 1.   The number of days you have gear among fish.
- 2.   The amount and/or effectiveness of gear.
- 3.   The fishing and carrying capacity of your boat or boats.
- 4.   The available, harvestable fish stock.

Only the available, harvestable stock is beyond your management
control. If you desire greater production (really a greater share of
available stock), you need only increase effort, gear, and/or boat
capacity. An increase in effort, gear, and/or boat capacity will increase costs. Therefore, the higher the production, the higher the
costs, *for any one season*.

Costs increase with production, but not in the same proportion.
The exact proportion or relationship will depend upon your management and the fishery. However, we can illustrate the general relationship by using a simple example. Suppose you own a clam boat and
are interested in harvesting Chesapeake Bay clams. During the com-
ing season you could take it easy and harvest 2,000 bushels, you
could put in an average effort and harvest 8,000 bushels, or you
could go all out and produce 14,000 bushels. Hypothetical total
costs at each of these and other production levels are shown in Table
6-1.

When production is zero, costs are $4,000 — the fixed cost.
When production is 2,000 bushels, costs are $14,000. When pro-
duction is 4,000, costs are $21,000, and so forth. Note that the dif-
ference in costs between 2,000 and 4,000 bushels is $7,000, the dif-
ference between 4,000 and 6,000 bushels is $5,000, and the differ-
ence between 6,000 and 8,000 bushels is $4,000. Also notice that
the difference in costs between 10,000 and 12,000 bushels is $7,000,

and between 12,000 and 14,000 bushels the difference is $21,000.
Not only are costs higher at higher production levels, they also become proportionately higher when the boat, crew, and gear approach
capacity.

This is more clearly seen in Figure 6-1, a graph of the production
and cost information in Table 6-1. Each point on the graph in
Figure 6-1 represents a production-cost combination from Table 6-1.
If we connect each point with a straight line, we approximate a clam
cost curve. The curve increases at a decreasing rate, and then at an
increasing rate at higher production levels; that is, it increases but becomes less steep, and then eventually becomes steeper. This general
production-cost relationship exists for any fishery. Its usefulness
will be apparent later in this chapter.

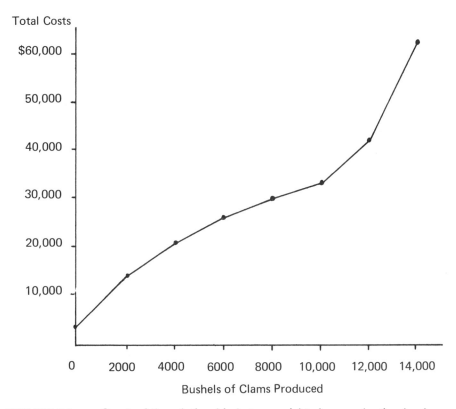

FIGURE 6-1.   Graph of the relationship between eight clam production levels
and total costs, given the season, gear, and boat size (hypothetical data).

Production and Returns

Economic theory tells us that gross returns increase with production. This is so obvious that it needs no elaboration, except for one problem. In our clam example you may be able to sell 2,000, 4,000, or even 6,000 bushels of clams for $5.50 per bushel, but to sell 10,000, 12,000, or 14,000 bushels you may have to lower your price to $4.70, $4.20, or $3.80 per bushel. (If you're wondering why, review Chapter 4.) When higher production levels can be sold only at reduced prices, gross returns do not increase proportionately with production.

This is illustrated in Table 6-2, where per-bushel price is $5.50 for 2,000, 4,000, 6,000, or 8,000 bushels. The price must be dropped to $4.70 to sell 10,000 bushels, to $4.20 to sell 12,000 bushels, and to $3.80 to sell 14,000 bushels. Gross returns are higher at higher production levels, but not proportionately so. This can be seen clearly in Figure 6-2, a graph of the information in Table 6-2. In Figure 6-2 gross returns increase at a constant rate to 8,000 bushels, at which point the rate decreases because of dropping prices. If 10,000, 12,000, or 14,000 bushels of clams could be sold at $5.50 per bushel, there would be a straight diagonal line across Figure 6-2.

Production and Profit

To observe the production-profit relationship, we need only bring the cost and returns relationships together. For our hypothetical clam-fishing business there is a profit or loss at each production level. This is found by subtracting costs from gross returns at each production level.

Table 6-3 lists the results of such subtractions, as well as some other useful information. We can see that if 8,000 or 10,000 bushels are produced, profits are greater than at any other production level. In fact, there is no profit if less than 4,000 bushels or more than 12,000 bushels are produced (given our eight observations). Profits increase from 4,000 to 10,000 bushels, then decrease.

In the last two columns of Table 6-3, we have calculated the increase in total costs and gross returns from one production level to the next. These were defined earlier as marginal costs and marginal returns. Notice that in moving from 8,000 to 10,000 bushels, the marginal cost is the same as marginal returns, and profit is maximum at $14,000. This is not a coincidence! Marginal costs will always

TABLE 6-2.   Relationship between eight clam production levels and gross
              returns (hypothetical data).

Bushels of Clams Produced	Price per Bushel	Gross Returns
0	$5.50	$ 0
2,000	5.50	11,000
4,000	5.50	22,000
6,000	5.50	33,000
8,000	5.50	44,000
10,000	4.70	47,000
12,000	4.20	50,400
14,000	3.80	53,200

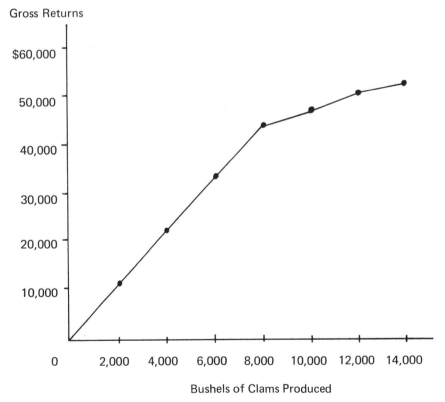

FIGURE 6-2.   Graph of the relationship between eight clam production levels
               and gross returns (hypothetical data).

TABLE 6-3.   Relationship among eight clam production levels, total costs,
gross returns, and profit (hypothetical data).

Bushels of Clams	Total Costs	Gross Returns	Profit	Increase in Total Costs (Marginal Costs)	Increase in Gross Returns (Marginal Returns)
0	4,000	0	-4,000		
				10,000	11,000
2,000	14,000	11,000	-3,000		
				7,000	11,000
4,000	21,000	22,000	1,000		
				5,000	11,000
6,000	26,000	33,000	7,000		
				4,000	11,000
8,000	30,000	44,000	14,000		
				3,000	3,000
10,000	33,000	47,000	14,000		
				9,000	3,400
12,000	42,000	50,400	8,400		
				21,000	2,800
14,000	63,000	53,200	-9,800		

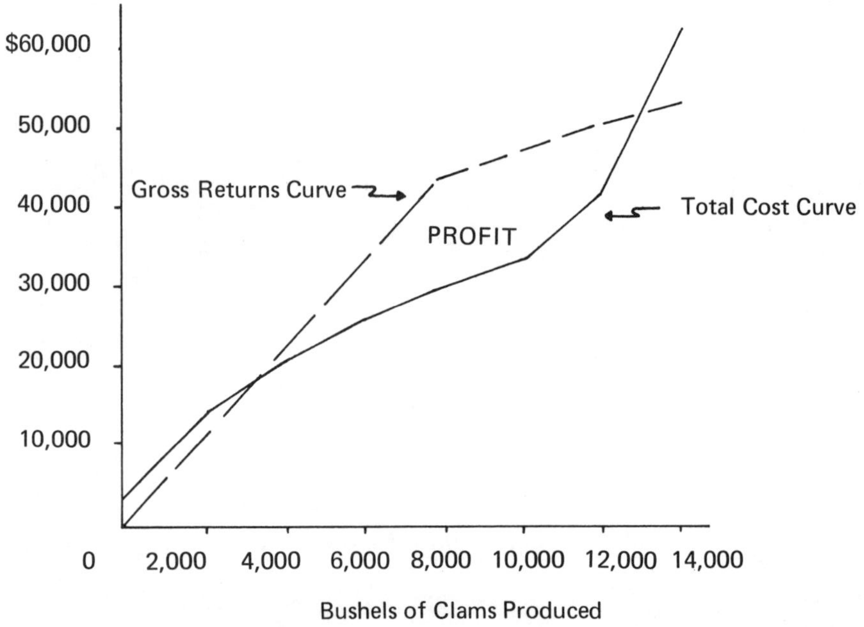

FIGURE 6-3.   Graph of the relationship among eight clam production levels,
total costs, gross returns, and profit (hypothetical data).

equal marginal returns at maximum profit. Although you will rarely
have sufficient data to calculate marginal costs and marginal returns,
it is still an extremely useful principle to remember. Fortunately, it
is also simple: *When profit is maximum, marginal costs equal marginal returns*.

Now that we have combined information from Tables 6-1 and
6-2, it is enlightening to combine Figures 6-1 and 6-2. This is accomplished in Figure 6-3. The solid line connects all the clam-pro-
duction/cost points and the dashed line connects all the clam-produc-
tion/gross-returns points. From Table 6-3, we see that profit is
minus $4,000 and minus $3,000 at zero and 2,000 bushels, respectively (a minus profit is a loss). In Figure 6-3, the cost curve is
above the gross returns curve at zero and 2,000 bushels. When
4,000, 6,000, 8,000, 10,000, or 12,000 bushels are produced, there
is a profit. Gross returns are above costs at each of these production
levels in Figure 6-3. This is the area of profit, as is marked on the
graph.

We can readily see from Figure 6-3 that profit can be made over
a wide range of production levels. We can also see that maximum
profit is not insignificant. Of course, our data is hypothetical and
the graph would look quite different if fixed costs were $10,000 in-
stead of $4,000, or if prices began at $4.50 per bushel instead of
$5.50, or if variable costs increased more rapidly. The effect of such
changes upon the maximum possible profit and the area of profitability is important, and is illustrated by the three graphs in Figures
6-4, 6-5, and 6-6.

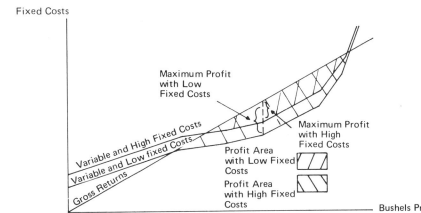

FIGURE 6-4.   Maximum profit and profit area with low and high fixed costs.

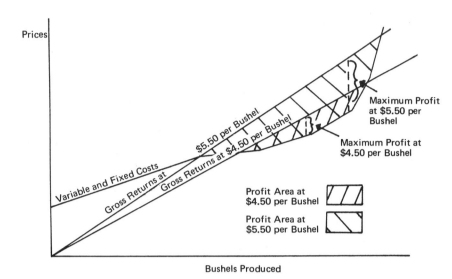

FIGURE 6-5.   Maximum profit and profit area with low and high prices.

These three graphs have been simplified by removing the numbers. Figure 6-4 illustrates the effect of higher (or lower) fixed costs. Variable costs and gross returns are not allowed to change in this illustration. The maximum possible profit and the profit area are reduced at higher fixed costs, because total costs are greater at each production level. Note, however, that maximum profit is attained at the same production level with both high and low fixed costs.

Higher or lower prices are illustrated in Figure 6-5. The gross returns line is steeper at $5.50 per bushel than at $4.50 per bushel. Again, the maximum possible profit and the profit area are greater at higher prices. Notice also that at a higher price, maximum profit is attained at higher production levels. The difference between Figure 6-4 and Figure 6-5 is significant. It tells us that if you are maximizing profit, a change in fixed costs doesn't necessarily require a change in production level in order for you to continue maximizing profit.

Figure 6-6 shows us that the maximum possible profit and the profit area are greater with lower variable costs. Because variable costs are related to production, the difference between the low- and high-cost curves is less at low production levels than at high production levels. The result is similar to, but opposite, the low-versus-high-price result in Figure 6-5; maximum profit is attained at lower

production levels with higher variable costs, and at higher production levels with lower variable costs.

Inputs, Costs, Returns, and Profit

The relationship among production, costs, returns, and profit is
informative and important. Nevertheless, fishermen seldom have the
opportunity to determine production. Most fishing business decisions
concern inputs, and inputs determine production. The important
production-determining inputs were listed earlier in this chapter.
They are: fishing effort, amount and/or effectiveness of gear, and
fishing and carrying capacity of the boat or boats. Decisions regarding
these inputs are illustrated in Chapters 13, 14, and 15.

In the remainder of this chapter, we will explain very briefly the
input-cost-returns-profit relationship. The profit-maximizing level
of any input is the level at which the increase in cost resulting from
an additional input is equal to the resulting increase in gross returns.
This is identical to the principle we identified earlier, except that
here we are deciding upon inputs and estimating the resulting pro-

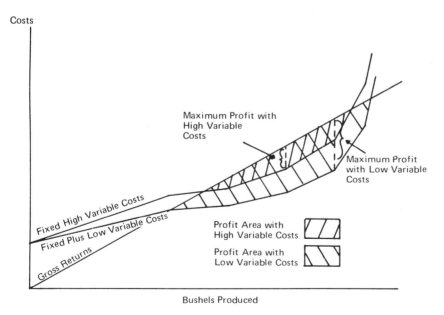

FIGURE 6-6.   Maximum profit and profit area with low and high variable
costs.

duction level, instead of deciding upon production levels and estimating resulting costs.

Inputs and Costs

The cost of an additional fishing day is the sum of the fuel, ice, and bait used, and crew-share, repair, and other costs resulting from that day's fishing. The cost of a larger boat is the increase in repairs, fuel consumption, insurance, depreciation, opportunity cost of investment, and so forth. The cost of an additional unit of gear is the additional crew required, depreciation, repairs, opportunity cost of investment, and so forth. None of these are difficult to estimate. Estimates of increased costs are illustrated in Chapter 10.

Inputs and Returns

The increase in costs resulting from an additional fishing day, a larger boat, or an additional crab pot is not difficult to estimate. The increase in gross returns resulting from an additional fishing day, a larger boat, or an additional crab pot may be little more than a guess in many fisheries. Nevertheless, an estimate of the increased gross returns is critical and must be made. Any estimate is better than none. Chapter 7 provides some guidance in making production estimates (see the section entitled "Using Probabilities" in that chapter).

Recognizing the variability of production given any level of input, we expect production (and gross returns) to be higher with more inputs than with fewer inputs. This relationship is referred to as a production function.

We know that production will not continue to increase beyond a certain point, if only one or two inputs are increased. For example, adding a third or fourth crewman on a shrimp boat would likely result in greater shrimp production. Adding a fifth, sixth, and even seventh man would also result in greater production, for night and day fishing would become possible. However, adding a fifteenth, sixteenth, or seventeenth crewman might actually reduce production, as the crewmen would begin to get in one another's way.

Inputs and Profit

As greater amounts of an input are used, profits will become larger provided that: (1) any profit is possible, (2) additional inputs in-

TABLE 6-4.   Costs, gross returns, and profit resulting from fishing ten different numbers of crab pots (hypothetical data).

Number of Crab Pots Fished	Annual Pot Costs	Annual Production (Pounds)	Gross Returns	Marginal Costs	Marginal Returns	Profit
100	$2,000	18,000	$ 9,000			$ 7,000
				$200	$2,000	
110	2,200	22,000	11,000			8,800
				200	1,500	
120	2,400	25,000	12,500			10,100
				200	1,000	
130	2,600	27,000	13,500			10,900
				200	500	
140	2,800	28,000	14,000			11,200
				200	200	
150	3,000	28,400	14,200			11,200
				200	100	
160	3,200	28,600	14,300			11,100
				200	100	
170	3,400	28,800	14,400			11,000
				200	0	
180	3,600	28,800	14,400			10,800
				200	0	
190	3,800	28,800	14,400			10,600

crease production, and (3) the noncontrollable production-determining factors are accounted for. However, the profit-maximizing input level is not where production is maximum; it is where marginal costs equal marginal returns.

This is clearly illustrated in Table 6-4, where we can observe the costs, returns, and profits resulting from fishing 100, 110, . . . , 190 crab pots. If 100 pots are fished, the annual costs of these 100 pots are $2,000, our best estimate of production is 18,000 pounds, and, at $.50 per pound, gross returns are $9,000. Profit for 100 pots is $7,000. With 110 pots, costs are $2,200, production 22,000 pounds, and gross returns $11,000 (at $.50 per pound). Fishing an additional 10 pots results in higher costs, higher production, and higher gross returns — until we reach 170 pots. At this point, there are more pots than we can haul effectively with our boat and many go unattended.

Fishing additional pots does not result in additional crabs; therefore, it is not rational to fish more than 170 pots. Production is maximized with 170 pots, but profit is maximized with 150 pots. In fishing 150 pots, the marginal costs ($200) equal the marginal returns ($200) and profit is $11,200. In fishing 170 pots, marginal costs do not equal marginal returns and profit is $11,000. Profit in this case is gross returns less pot costs only.

Few managers would have the kind of data illustrated in Table 6-4 and would be able to determine the profit maximizing number of pots so precisely. Nevertheless, the principle involved is useful regardless of the accuracy of your information: the effort, size of boat, units of gear, and so forth, that maximize production do not necessarily maximize profit; profit is maximum when the increase in gross returns (marginal returns) equals the added cost of additional effort, a larger boat, more gear, and so forth (marginal costs).

7 COPING WITH RISK
AND UNCERTAINTY

"The only things we can be certain of are death and taxes, and even death is not so certain any more." Every businessman must make decisions with incomplete information, even though those decisions involve some risk or uncertainty as to the outcome. This is especially true in commercial fishing.

The tools of economics and statistics can help you incorporate considerations of risk and uncertainty in fishing business decision making. I have not attempted to discuss here all the intricacies of economics and statistics, but I have presented some definitions and illustrations that will enable you to use what information you have to improve your chances of a profitable trip or year.

Certainty, Risk, and Uncertainty Defined

Certainty exists where the outcome of some decision or action can be exactly predicted — and that outcome always occurs. We can predict exactly when the sun will rise each day, and that prediction becomes true without fail. This is certainty.

Risk exists when the outcome of some decision or action cannot be predicted exactly, but can be predicted with some degree of possibility or probability. When the outcome occurs, it may or may not be as predicted. The National Weather Service may predict 20-knot winds out of the northwest for tomorrow, knowing that there is a 70 percent chance that a weather system may continue its present course and cause those winds. However, the weather system may actually stall and tomorrow might be perfectly calm. In this case, the 3-out-of-10 chance rather than the 7-out-of-10 chance occurs.

Uncertainty exists when the outcome of some decision or action cannot be predicted and when the outcome could be just about anything. This is a situation where anybody's guess has equal value; where the outcome could range from the "ridiculous to the sublime." You know that there is a chance that you might be rammed and sunk while drifting in the open sea at night, but you don't know what that chance is. You can say that it's unlikely to happen, but you can't make a prediction or probability statement about it. This is uncertainty.

Risk is the situation most of us face most of the time. Even uncertainty can be turned into a risk situation as we gain knowledge and experience. Of course, some situations have more risk than others; that is, some outcomes can be predicted more confidently than others: insurance costs can be predicted with great confidence, annual landings with less.

As a manager you can use predictability or probability information in decision making, or you can simply protect yourself from the adverse effects of risk and uncertainty situations.

Decision Making in the Face of Risk

When faced with a risk situation, you may respond in one of the following ways:

1. You may do nothing, feeling that you must have absolute certainty before making a decision. "I ain't gonna do nuthin' 'til I'm sure." This type of fisherman was characterized in Chapter 2 as the person who may be happier working for someone else. He is not a decision maker and prefers to leave all the risk to the boss.

2. You may close your eyes and plunge ahead, ignoring any probability information. You always assume that "things will work out OK," and you don't want to be confused with the facts. This type of fisherman must have a gambling spirit and little to lose.

3. You may look at several courses of action and choose the one whose worst possible outcome is not as bad as the worst possible outcome for each of the other alternatives. We can call this the "pessimist" decision criteria.

   Assume that you have the choice of going fishing facing 40-knot winds and 8-foot seas or staying home and repairing gear. If you fish, the worst thing that could happen is that you would wreck your boat and perhaps yourself. If you stay home, the worst thing that could happen is that you would give up $8,000 worth of fish. Staying in port has the least detrimental outcome, so you stay in port.

The fisherman with a valuable boat, high equity in the business, and a personal dislike for taking chances, and who is close to retirement, will frequently use the "pessimist" criteria. Why take a chance when there is so much to lose?

4. You may look at several courses of action and choose the one that has the best of all the best possible outcomes. We can call this the "optimist" decision criteria.

    Using the above example, the "optimist" decision maker would see that fishing in spite of 40-knot winds and 8-foot seas might bring in $8,000 whereas working on gear in port might save only $300. This fisherman looks on the bright side of things and goes fishing. He probably has fewer personal responsibilities than the "pessimist" fisherman, little equity in his boat, more of a gambling spirit, and expects to "make it big" one day.

5. Finally, you may look at several courses of action, determine the chances (probability) of each outcome occurring, and play the odds. This would be the "most-likely" decision criteria. You would be using all the risk information available, in contrast with the first two decision makers above, who use no risk information.

Decision Criteria Illustrated

There is considerable risk in selecting a fishery. We can draw upon some research conducted by the NMFS and illustrate this risk. Figure 7-1 shows the maximum and minimum profits of a sample of three fishing businesses, one operating in the Gulf of Mexico shrimp fishery, one in the North Pacific halibut fishery, and the third in the New England scallop fishery.

The minimum profit experienced over 7 years of shrimp fishing was −$3,200, and the maximum profit was $8,100. The minimum profit experienced over 3 years of halibut fishing was $4,000, maximum $6,500. The minimum and maximum for 7 years of scallop fishing were −$1,500 and $7,700, respectively.

A decision using the "pessimist" criteria would result in entering the halibut fishery, for the worst profit experienced in that fishery was $4,000, which is greater than the worst profit for shrimp (−$3,200) or for scallops (−$1,500).

TABLE 7-1.   Profits experienced in shrimp, halibut, and scallop fishing, and average profit for each (NMFS data).

Observation	Shrimp	Halibut	Scallops
1	$- 3,200	$4,000	$- 1,500
2	1,500	4,900	- 900
3	2,700	6,500	2,900
4	5,200		3,000
5	5,900		3,000
6	6,600		6,300
7	8,100		7,700
Average	$ 3,829	$5,133	$ 2,929

A decision using the "optimist" criteria would result in shrimp fishing, for the best profit experienced in that fishery, $8,100, was better than that in the halibut ($6,500) and scallop ($7,700) fisheries. Both of the above decisions assume that the past predicts the future.

The third decision criteria requires some knowledge of probability. For example, how many years was shrimp profit negative, how many years was it positive, and what was the average profit compared with scallop profit? Was the worst or best profit for each of the three far below or far above the average for each?

Table 7-1 will help us to answer these questions and to make our decision on the basis of all the risk information available. We can see that, although shrimp fishing had the highest possible profit and halibut the highest lowest profit, halibut also had the best average profit. However, our information on halibut is based on only three observations, whereas our shrimp and scallop information is based on seven observations. We can have more confidence in the shrimp and scal-

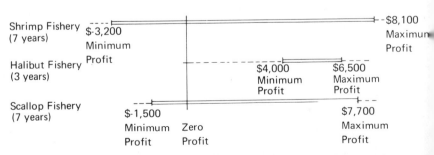

FIGURE 7-1.   Chart of the maximum and minimum profits experienced in shrimp, halibut, and scallop fishing (NMFS data).

lop information. We are not able to predict future halibut profits as
accurately as future shrimp and scallop profits. Therefore, unless we
can obtain more information on halibut profits, we may decide to
choose between shrimp and scallops.

Let's study the figures in Table 7-1 once more. We can see that
there are four observations of shrimp profits over $5,000, one for
halibut over $5,000, and two for scallops over $5,000. We can see
that the average profit for shrimp is lower than the average profit for
halibut because one observation was very low (−$3,200). In spite of
the two averages, it is clear that the figures show that you would
be more likely to earn a larger profit shrimp fishing than halibut fishing. This is more clearly shown by comparing the three graphs in
Figure 7-2. The base line in each of these graphs is identical to the

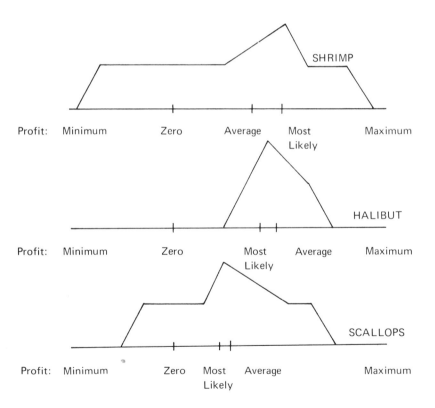

FIGURE 7-2.   Graphs illustrating the likelihood of different profits for
shrimp, halibut, and scallop fishing (based on NMFS data).

corresponding base line in Figure 7-1. The height of the curve above the base line is an index of the likelihood (probability) that that level of profit will occur: the higher the curve, the greater the probability of obtaining that profit.

Notice that the most likely profit for shrimp is greater than the most likely profit for halibut, which, in turn, is greater than the most likely profit for scallops. Also, the most likely profit for shrimp is greater than its average profit, but the most likely profits for halibut and scallops are less than their average profits.

Combining all this information and assuming that the past predicts the future accurately, the decision maker who chooses the shrimp fishery will experience the highest profit over time.

Using Probabilities

With an increasing amount of general fishery information becoming available and with continual improvements in your own accounting and fishing information, you will increasingly be able to use probability estimates in making management decisions. The following paragraphs describe a useful procedure for using probability estimates.

Assume you are lining up a crew for a halibut schooner and must guarantee a minimum of $5,000 for the season to each crew member. The size of the crew is related to the amount of gear and the landings. As illustrated in Table 7-2, if you have a crew of 3, your crew costs are at least $15,000. If you increase crew size to 4, crew costs go up $5,000 — from $15,000 to $20,000. An increase from 4 to 5 crewmen increases costs another $5,000, and so on.

Under conditions of certainty we can predict exactly what will happen to gross returns. (Certainty, of course, doesn't exist in the halibut fishery, but assuming it does will help us make an important point.) Increasing crew size from 3 to 4 increases gross returns by $7,000 under certainty conditions. Does it pay to increase crew size from 3 to 4? Five thousand dollars are added to costs, but $7,000 to gross returns — an addition to profit of $2,000. It pays!

Will it pay to add another crewman? Five thousand dollars will again be added to costs, but this time only $5,000 instead of $7,000 are added to gross returns. There is no addition to profit.

Suppose we add the sixth crewman. Costs again go up by $5,000, but gross returns go up by only $4,000. Adding the sixth man re-

LE 7-2.   Crew size, costs, and returns in halibut fishing, illustrating marginal returns under conditions of certainty, risk, and uncertainty (hypothetical data).

mber of wmen	Crew Costs	Marginal Cost (Addition to Crew Costs)	Marginal Returns Under Certainty (Addition to Gross Returns)	Marginal Returns Under Risk (Probability Estimates of Additions to Gross Returns)					Marginal Returns Under Uncertainty (Addition to Gross Returns)
				.10	.20	.40	.20	.10	
3	$15,000								
		$5,000	$7,000	$5,000	$7,000	$9,000	$11,000	$13,000	$0-?-13,000
4	20,000								
		5,000	5,000	4,000	5,000	7,000	9,000	11,000	0-?-13,000
5	25,000								
		5,000	4,000	3,000	4,000	5,000	7,000	9,000	0-?-13,000
6	30,000								
		5,000	3,000	2,000	3,000	4,000	5,000	7,000	0-?-13,000
7	35,000								
		5,000	2,000	1,000	2,000	3,000	4,000	5,000	0-?-13,000
8	40,000								

duced total profit by $1,000. Under conditions of certainty the decision is easy: a 5-man crew will maximize profit.

Now let's look at the more realistic risk situation. If crew size is increased from 3 to 4, costs increase $5,000 and there is a 10 percent probability of gross returns increasing by $5,000, a 20 percent probability of gross returns increasing by $7,000, a 40 percent probability of gross returns increasing by $9,000, a 20 percent probability of gross returns increasing by $11,000, and, finally, a 10 percent probability of gross returns increasing by $13,000. Given this probability information, would you add the fourth man to your crew?

The "pessimist" decision maker sees that under the worst conditions, nothing is gained or lost — so he would add the fourth man. The "optimist" decision maker sees the 10 percent probability of a $13,000 increase in gross returns and would add the fourth man, expecting an $8,000 increase in profit.

The above decisions are reasonable; however, neither of them uses all the probability information available in our example. Because we know the probability of each outcome, we can multiply each probability by its corresponding outcome and add the results, thereby obtaining the exact mathematical expectation for each outcome. For example, the expected increase in gross returns as crew size goes from 3 to 4 is calculated as follows:

$$(.10 \times \$5,000) + (.20 \times \$7,000) + (.40 \times \$9,000) + (.20 \times \$11,000) + (.10 \times \$13,000) = \$9,000$$

The most likely increase in gross returns ($9,000) is greater than the $5,000 increase in cost — so the fourth crewman would be added.

What about adding the fifth crewman? The "pessimist" decision maker would not add the fifth man because he would see that under the worst situation the gross returns increase is less than the cost increase ($4,000 < $5,000). The "optimist" decision maker would add the fifth man, for under the best situation the $11,000 increase in gross returns is greater than the $5,000 increase in costs.

The exact-probability decision maker finds that

$$(.10 \times \$4,000) + (.20 \times \$5,000) + (.40 \times \$7,000) + (.20 \times \$9,000) + (.10 \times \$11,000) = \$7,100$$

which is greater than the cost increase — so he also would add the fifth crewman. He would add the sixth crewman, as would the "optimist" decision maker, but would not add the seventh because

$$(.10 \times \$2,000) + (.20 \times \$3,000) + (.40 \times \$4,000) + (.20 \times \$5,000) + (.10 \times \$7,000) = \$4,100$$

which is less than the cost increase.

The "optimist" decision maker would continue to add the seventh and eighth crewman because he always expects the best outcome, regardless of probability.

Suppose that you know only that the increase in gross returns would be somewhere between zero and $13,000, but that you had no probability estimates. This would constitute uncertainty and would be very difficult circumstances in which to make a decision. It would become necessary for you to act on the basis of an assumption. Suppose that you assume that the increase will be greater than $5,000, and that you add the fourth crewman and gain a year's experience with him. This results in having more information than before; it moves you away from uncertainty and toward risk.

Protection Against Risk and Uncertainty

Risk and uncertainty are a part of fishing. Even though you may use all the available information and choose the least risky alternative, the amount of risk or uncertainty remaining may still be unacceptable to you. The probability of it happening may be small, but the loss of your vessel might put you out of business. Injury to a crewman and a liability suit may bankrupt you. Two or three bad years may result in the foreclosure of your loans and force you to look for another job.

You can obtain protection against these and other risks. In fact, you can obtain some protection against uncertainty by first joining others sharing the same uncertainty. Uncertainty for one becomes risk for a group. For example, you cannot predict when one boat will burn, but you can predict that one boat of a group of 125 will burn each year. However, protection against risk and uncertainty comes at a cost, sometimes a very high cost. You must weigh the cost of this protection against the potential financial loss of a sunk vessel, a lawsuit, a foreclosed loan, and so forth.

The following are some common methods of protecting against risk and uncertainty:

Crew-share

Nearly every commercial fisherman pays his crew on a percentage basis. The gross returns and some costs are shared among the crew, skipper, and owner. This reduces the skipper's and owner's risk from a poor trip at the cost of reducing the potential gain from a good trip. Various percentages or lay systems allocate more or less risk to the crew.

Insurance

An insurance premium is what you pay someone else to accept a percentage of risk damage or to accept any risk damage over an agreed-upon limit. There are several kinds of insurance:

1. *Property insurance* protects you against loss or damage to vessel, gear, truck, home, and other property. Hull insurance is common among larger fishing vessels. Gear insurance is rarely purchased, however: most fishermen are willing to accept the risk and bear the complete cost of replacing nets, traps, lines, doors, and so on.

2. *Liability insurance* protects you against liability claims. As an employer and property owner you can be held liable for injury to employees or damage to other property and,

through civil action, can be required to compensate others for such injury or damage. In the fishing industry, this is called protection and indemnity (P & I) insurance. It can be costly if you fish in the open ocean because admiralty law places more responsibility on the skipper of an ocean-going vessel than on the operator of a land-bound business.

3. *Life insurance* protects your family and your business against the financial burden of your untimely demise. It is an important kind of business insurance.

4. *Health and accident insurance* can protect you against excessive costs of regaining your productive capacities or against the loss of income while you are incapacitated. It can be important for you because physical abilities are critical in operating a fishing business. It is hard to catch fish from a hospital bed! However, health insurance companies recognize the greater risk of injury in fishing, and the premiums are therefore higher for fishing than for many other occupations.

5. *Income insurance*, though uncommon, is available; it compensates you for low-income seasons. This is accomplished by your paying large premiums during high-income seasons.

Diversification

If you participate in three or four different fisheries, you are likely to experience less income fluctuation than the fisherman who concentrates on one fishery. In the Pacific Northwest, it is common for fishermen to troll for salmon from June to August, troll for tuna during August and September, and pot-fish for crab from December to April. If there is a failure in one or even two of these fisheries, there is still some protection of income in the other one or two fisheries. However, if all the fisheries in which you participate fail, risk will come home in full force!

Diversification is done at a cost: a greater investment in gear, a less efficient vessel for one or more of your fisheries, gear changeover time, and greater skill and management requirements.

Flexibility

This is closely related to diversification. If you are able to move from one fishery to another with minimum effort and cost, you can take advantage of the good fisheries in any one season and avoid the poor ones. This requires geographic mobility, a wide range of skills, and a flexible vessel. The cost is in the time devoted to moving from one fishery to another and in not having the skills, vessel, and gear to be really efficient in one or two of the fisheries. Also, there is a greater chance of making wrong decisions — of changing to one declining fishery after another.

Market Arrangements

You can obtain guaranteed prices for fish and fishing supplies to protect yourself against market risk. However, by protecting yourself against lower fish prices and higher supply prices, you give up the possible windfall gains from higher fish prices and lower supply prices. Another way to protect yourself from market risk is to participate in the market process itself. You can join a cooperative and benefit from its successes, you can buy stock in an incorporated sea-food marketing operation and benefit from its successes (or failures!), or you can enter the sea-food marketing business yourself. This is called vertical integration. The fisherman who processes and markets his own fish is also diversifying. However, all these market arrangements require an investment of your capital in a nonfishing enterprise, and most commercial fishermen have enough difficulty obtaining sufficient capital to keep their fishing business efficient! This leads us to the last and probably most important method of protection against risk and uncertainty.

Financial Management

If you are doing everything possible to maximize your profits, you are protecting yourself against risk and uncertainty. The fisherman with $10,000 in the bank and 80 percent equity (ownership) in his business can withstand more disasters than the fisherman with $10 in the bank and a $10,000 annual payment on a vessel loan. Keeping your assets in a form that can be readily converted to cash and keeping fixed payments low provide you room to maneuver around disasters. Note repayment terms that are negotiable annually or are a percentage of gross returns protect you against foreclosures in a risky fishing business. Investing in readily marketable gear and electronics may allow you to convert these assets to cash quickly to cover unexpected costs, instead of having to sacrifice fishing capabilities. This means standard instead of experimental otter doors, traps, gurdies, and so forth, and portable radios and radar instead of built-in equipment.

Risk and uncertainty are very much a part of the fishing business. Part of being a fisherman is accepting this. Risk and uncertainty imply potential gains as well as losses, a point often forgotten. By using the decision criteria outlined in this chapter, you can take account of the risk and uncertainty inherent in your fishing business, and through crew-shares, insurance, diversification, flexibility, market arrangements, and financial management, you can protect yourself against them.

PART TWO

MANAGEMENT TOOLS

8 DEVELOPING MANAGEMENT INFORMATION

The economic principles and management tools discussed in this book are of little value without good management information. Fishermen frequently ask, "What kind of records should I keep?" In this chapter, we will illustrate some record systems and discuss some record-keeping principles.

Management Information Requirements

The ideal record system should have the following characteristics:
1. It should be a system you have developed for yourself and with which you are intimately familiar.
2. It should be based primarily upon your own management needs and, second, upon income tax reporting requirements.
3. It should be simple and flexible yet provide necessary information readily.

Perhaps the best way to start is with a blank sheet of paper. List your information needs and then develop a system that will satisfy those needs simply and efficiently. Some of the information needs to be considered are the following:

Profit (Returns to Labor, Management, and Investment) from your Various Fishing Activities

This may require a record of costs attributable to each fishery, the investment in each fishery, a log of your effort in each fishery, and gross returns.

Efficiency of your Fishing Operations

This may require a detailed log of your time, vessel time, and crew time, and a record of costs and returns by major category.

Costs would include gear repair costs, vessel repair costs, fuel costs, and ice and bait costs. Also required is a log of weight and value of fish landed each day, week, or trip.

Financial Management of your Business

This requires a listing of accounts payable, accounts receivable, and assets (a net worth statement). Also required is a cash-flow budget showing when receipts are expected and when major payments will be made.

Development of Fishing Skills

Much can be learned from a detailed log showing quality and amount of fish caught, gear used, how used, oceanographic conditions, location, and weather conditions.

Government Reporting Requirements

This is the least important information needed for fishing business management. This requirement can usually be met with very little additional effort if the preceding four needs are satisfied. You may also have to keep some record of income tax withheld, social security taxes withheld, and other withholdings from the crew's pay. Also, an inventory of assets may be required in any estate settlement proceedings.

Some Accounting Concepts

You may find the assistance of a professional accountant helpful in developing management information. In addition to accountants, there are other sources of assistance: banks, computerized accounting services, and specialized business services. Whichever you choose, you should check their competence and be sure you are getting your money's worth. An investment in accounting services should be viewed as a source of profit. You should analyze this investment as you would a boat or gear investment.

Most accountants and accounting service agencies have their own jargon. Understanding some of the accounting jargon and concepts will help you in developing your own information system as well as in talking with accountants.

Accounting procedures have been developed to serve those who wish to pay for such services. Manufacturing firms, wholesale and retail businesses, and service firms almost always hire accountants; fisherman, farmers, miners, and loggers hire accountants only occasionally. Therefore, standard accounting procedures and terminology do not always fit the fishing business. For example, most accountants use a double-entry system, which means that for every debit entry there must be an offsetting credit entry in the books, and vice versa. This approach helps "balance" the books: it helps account for every dollar in the business, an important consideration for stock corporations.

Another procedure used by accountants is the trial balance. This is simply a timely summation of debits and credits to make sure that they are the same. The chart of accounts is the categorization of all expected expense (debit) and income (credit) transactions. By establishing a category for each potential transaction at the beginning of the accounting period, the proliferation of inconsequential categories is minimized.

The cash and accrual bases of accounting constitute another important concept. If you operate on a cash basis, you treat each expense as a debit when it is paid and each income as a credit when it is received. At the end of each trip, you owe the crew their share and the processor owes you for the fish he purchased; but the crew-share becomes an expense three days later when you pay the crew, and the amount owed you by the processor becomes income when you receive his check the next week. On the accrual basis, you treat the expense as a debit when it is incurred and the income as a credit when it is owed. On this basis, the moment you complete your trip, the crew-share becomes an expense and the fish delivered to the processor becomes income.

Record-Keeping Procedures

The following are some procedures that will help you establish and maintain a complete management information system.

Separate Checking Accounts

An absolute minimum for any management information system is to establish two checking accounts. One should be for the business and one for personal or family use. Do not use the family checking account for the fishing business. With separate accounts, much of the record keeping is left to the bank. *All* fishing business income must be deposited in the business checking account and *all* business expenses must be paid by check out of that account. This may mean estimating petty-cash needs and writing a check each week to cover these needs. Beyond this, all expenses should be paid by individual check, not with cash. Funds for budgeted family living expenses can be transferred to the family or personal checking account monthly or weekly. Withdrawals from the business for personal expenses can more easily be controlled with separate accounts, but most important, all business transactions will be recorded at least once — in your business checkbook.

Diaries or Logbooks

If an accurate and complete business checking account is basic and essential in record keeping, a complete diary and/or logbook is a close second in importance. The form or type of diary and/or log depends upon personal convenience and eventual use. A small notebook that can be carried in a jacket or shirt pocket works well. A bright, waterproof cover increases visibility and durability. An attached pencil reduces the likelihood of being without that handy item.

You should enter all information that has the slightest possibility of being useful. As you gain experience in using your diary and/or log, you can be more selective in what you enter. It is a good place to record business trips (place, purpose, and date), names and addresses, technical data on gear and electronics, fishing information, market information, indicators of boat and gear performance, and other management information. An easy-to-use, sturdy camera and cassette tape recorder can be used to supplement the diary and logbook information. Photographs and tape recordings are efficient substitutes for handwritten diary and logbook entries.

Permanent Ledgers

The third most important part of a management information system is a set of permanent ledgers. As indicated earlier, the best ledgers are those you develop for your own management needs. However, examples of some standard ledger forms may help you in organizing **your** own system. Figures 8-1 and 8-2 illustrate a basic ledger for recording receipts and expenses. These ledgers should be kept in a permanent, safe location and should be updated at least once per month. One ledger page can be used for each week, each month, or each fishery. You can also use one page of expenses per month for each fishery. This will require more effort in recording information but will save time when you analyze the results.

If you buy something that you expect to last more than one year, it should not appear in the expense ledger but should be treated as a capital purchase. When a hydraulic winch is purchased, for example, it is capitalized and placed on a depreciation schedule based on its expected life. The annual depreciation is then treated as the expense, which appears on its own ledger page. When the winch is sold, it is treated as a capital sale, not as an ordinary receipt. Figure 8-3 illustrates a ledger form for capital purchases and sales.

Another useful part of your management information system is a net worth statement. This lists your assets and liabilities and tells you how much you are worth at a given point in time. A net worth statement is illustrated in Figure 8-4. Assets are what you own or what is owed you. Liabilities are what you owe others. Assets and liabilities can be classified in any convenient way, but are usually listed as shown in Figure 8-4. Assets less liabilities equals your net worth, or what you would have before taxes if you sold out completely, collected all accounts receivable, and payed all debts.

Although a net worth statement is usually prepared once a year, it is useful to examine your net worth more frequently. It is an indication of the financial progress of your business. The net worth statement (or balance sheet) is discussed in more detail in Chapter 9 under the heading, "Financial Analysis."

Depreciation

Depreciation is a technique for distributing the cost of a capital purchase over the life of the item. The amount of depreciation depends upon the initial investment and the capital asset's loss of value over time. Loss of value depends upon "wear and tear" and obsolescence. Depreciation on your gear shed is more a function of obsolescence, whereas wear and tear is the primary cause of depreciation on cables, winches, engines, otter boards, and hydraulic systems.

For management purposes, each capital asset should be depreciated according to its actual decrease in value. This may be 20 per-

Line	Total Receipts		Date	From Whom	Description of Item		Unit
	Cash	To be received				Quantity or Weight	(lbs.) (tons)
	$	$					

FIGURE 8-1.   Illustration of a simple receipt ledger for a fishing business.

Line	Total Expenses		Date	To Whom	Description of Item	Quantity or Weight	Unit (lbs.) (tons)	Price per unit
	Cash	To be paid						
	$	$						$

FIGURE 8-2.   Illustration of a simple expense ledger for a fishing business.

cent one year, 10 percent the next, and 50 percent the third. In practice, some standard method is usually applied to each asset. Three common methods are illustrated below.

The Straight-Line Method

In this method, the annual depreciation is computed by dividing the original cost, less any salvage value, by the expected years of life.

Price per unit	Salmon	Tuna	Crab	Shrimp	Flatfish	Rockfish		To be received
	$	$	$	$	$	$		

Crew-share	Vessel Repair		Gear Repair		Fuel	Food	Ice and Bait	Interest and Insurance	Licenses and Taxes	Misc.
	Labor	Material	Labor	Material						
	$	$	$	$	$	$	$	$	$	$

A $5,000 engine with a $500 salvage value and an expected life of ten years would have an annual depreciation of $450.

$$\$450 = (\$5,000 - \$500) \div 10 \text{ years}$$

The Declining-Balance Method

A fixed *rate* of depreciation is applied to the remaining value of the asset. There is no salvage value, for this is built in. The rate frequently used is double the straight-line rate. For example, the $5,000

CAPITAL PURCHASES

(1)	(2)	(3)	(4)	(5)	(6)	(7)	(8)	(9)
Date	Description	Cost Basis	Estimated Life	Salvage Value	Depreciation Method	Deprec. Value Beginning This Year	Deprec. This Year	Deprec. Value Ending This Year

TOTAL COLUMNS 3, 7, 8, and 9

FIGURE 8-3.   Illustration of a simple capital purchase and sale ledger for a fishing business.

engine depreciated for 10 years by the straight-line method yields a 10 percent rate of depreciation. Therefore, 20 percent would be a reasonable rate to use in the declining-balance method.

	Value at Beginning of Year	Annual Depreciation		Remaining Balance
Year 1	$5,000	(20% × $5,000) =	$1,000	$4,000
Year 2	4,000	(20% × 4,000) =	800	3,200
Year 3	3,200	(20% × 3,200) =	640	2,560
Year 4	2,560	(20% × 2,560) =	512	2,048
Year 5	2,048	(20% × 2,048) =	410	1,638

Year 6 and following: continue to take 20% of remaining balance

The Sum-of-the-Years'-Digits Method

The annual depreciation is computed by multiplying a fraction by cost less salvage value. The fraction for each year is the sum of the years of life divided into the years of life remaining. If the life is 10 years, the denominator (divisor) would be $1 + 2 + 3 + 4 + 5 + 6 + 7 + 8 + 9 + 10 = 55$, and the numerator would be 10. The fraction for the first of the 10 years is 10/55, for the second year, when 9 years of life remain, 9/55, and so forth. Using the same $5,000 engine with a $500 salvage value, the depreciation schedule for the first four years would look like this:

CAPITAL SALES

(10) Date	(11) Description	(12) Sales Price	(13) Cost Basis	(14) Depre- ciated Value	(15) Capital Gain or Loss	(16) Ordinary Gain or Loss

TOTAL COLUMNS 12, 15, and 16

	Value at Beginning of Year	Annual Depreciation	Remaining Balance
Year 1	$5,000.00	10/55 X ($5,000−$500) = $818.18	$5,000.00−$818.18 = $4,181.82
Year 2	4,181.82	9/55 X (5,000− 500) = 736.36	4,181.82− 736.36 = 3,445.46
Year 3	3,445.46	8/55 X (5,000− 500) = 654.54	3,445.46− 654.54 = 2,790.92
Year 4	2,790.92	7/55 X (5,000− 500) = 572.72	2,790.92− 572.72 = 2,218.20

The depreciation method that matches the asset value loss most accurately should be used. For example, your new pickup truck decreases in value rapidly the first two or three years of ownership, but less afterward. The declining-balance and sum-of-the-years'-digits methods would be most accurate for a new pickup truck.

External Management Information

Although it is extremely important for you to develop and maintain management information about your own fishing business, many of your decisions will be based in part on information from other fishermen, processors, equipment suppliers, magazines, newspapers, extension people, researchers, and government agencies. There are several things you can do to obtain information from these sources and to keep the information in a useful form.

Information gathered from conversation and observation along the waterfront should be recorded in your pocket diary or log. Names and phone numbers of people, dimensions and costs of gear,

Year Ending_____

(1) Assets	(2) Value, Beginning of Year	(3) Value, End of Year
Cash		
Accounts Receivable		
Vessel		
Gear		
Vehicles		
Structures		
Total Assets		
Total Liabilities from Columns 5 and 6		
Net Worth (Assets Less Liabilities)		

FIGURE 8-4.   Illustration of a net worth ledger for a fishing business. Only business assets and liabilities constitute business net worth. Both personal and business assets and liabilities constitute personal net worth.

and success and failure stories are all potentially useful at a later date. As indicated earlier, a small, easy-to-use camera can be valuable in recording fishing innovations. A portable tape recorder is useful in collecting ideas and helpful hints from conversations and meetings.

It may not pay to subscribe to every industry publication, but if you and several of your friends subscribe to different ones and circulate them, you can learn more at less cost. It may be useful to encourage local libraries to subscribe to industry publications and

(4) Liabilities	(5) Value, Beginning of Year	(6) Value, End of Year
Accounts Payable		
Mortgages		
Notes		
Total Liabilities		

to institute check-out privileges for them; or you may wish to establish a library among your own group.

It is not difficult to place yourself on the mailing lists of universities and state and federal agencies that disseminate fisheries information. Some of these are suggested in the bibliography. An inexpensive two- or three-drawer file makes it easier to store and retrieve published information. Some materials should be kept in the wheelhouse or on the pickup seat in a three-ring binder, clip board, or small brief case.

Participation in formal and informal educational programs sponsored by your association, community college, university extension program, or a state or federal agency can be an excellent source of management information, and frequently leads to other sources.

9 ANALYZING
THE FISHING BUSINESS

Monitoring the fishing business for profit, financial condition, and efficiency is an important part of fishing business management. It is the important step between keeping good records and making business decisions.

In this chapter we will illustrate profit analysis, financial analysis, and efficiency analysis. We will also demonstrate how you can use these tools to understand and improve your business.

Profit Analysis

In Chapter 5 we defined costs, returns, and profit. In Chapter 6 we saw how profit relates to costs and production. In this chapter we will examine profit "after the fact" — that is, after it has been earned. Let's illustrate by studying some good logbook and record information on a 40-foot combination troller and dungeness-crab boat. This information is presented in Table 9-1.

Profit is usually found by subtracting total costs from gross returns. "Total costs" was defined in Chapter 5 as variable costs plus fixed costs. Profit must therefore be:

$$\$40,900 - (\$21,685 + \$8,120) = \$11,095$$

| Gross returns | Variable costs | Fixed costs | Profit |

Although this is a common method of calculating profit, you will recall from our discussion in Chapter 5 that the resulting figure does not account for opportunity costs and does not provide a good year-to-year or business-to-business comparison.

TABLE 9-1.   Description, costs, and returns information on a 40-foot combination boat.

Vessel Description:

    40 feet by 14 feet, $60,000 market value, 12-ton capacity, refrigeration, and 200 crab pots

Fishing effort:

    65 days crab fishing, 40 days salmon trolling, and 55 days tuna trolling

Landings and prices:

    15 tons dungeness crab at $900 per ton, 12 tons salmon at $1,200 per ton, and 25 tons albacore tuna at $520 per ton

Variable costs: (explained in Chapter 5)

Repairs	$ 5,870
Food	1,720
Fuel	1,430
Transportation	1,200
Bait	990
Ice	90
Miscellaneous	160
Crew-share	10,225
TOTAL	$21,685

Fixed costs: (explained in Chapter 5)

Insurance	$2,300
Interest	2,200
Depreciation	3,000
Moorage	170
Licenses	110
Miscellaneous	340
TOTAL	$8,120

Gross returns:

15 tons crab at $900 per ton	$13,500
12 tons salmon at $1,200 per ton	14,400
25 tons tuna at $520 per ton	13,000
TOTAL	$40,900

The value of any unpaid family labor is one of the opportunity costs. Your wife, son, or daughter may invest considerable time in keeping books, chipping paint, building crab pots, cooking for the crew, and so forth. If you do not pay them, it will not appear as a cost in your business.[1] This may make your operation appear more profitable than your friend's. If your friend is a bachelor, the difference in profit would be a result of married status, not your fishing business management!

Return to Labor, Management, and Equity

A good diary or log will tell you how much time was contributed by family members. For our illustration, let's assume that the family contributed 160 hours of work on crab pots and 40 hours of bookkeeping. Assuming that gear workers normally receive $2 per hour and bookkeepers, $3 per hour, we must add the following to costs:

$$160 \text{ hours} \times \$2.00 = \$320$$
$$40 \text{ hours} \times \$3.00 = \underline{\$120}$$
$$\text{Value of unpaid family labor:} \quad \$440$$

We now have gross returns less variable costs, fixed costs, and value of unpaid family labor.

$$\$40,900 - (\$21,685 + \$8,120 + \$440) = \$10,655$$

Gross returns	Variable costs	Fixed costs	Family labor

The resulting figure is the amount left over to cover all other inputs to the business. It is the return to the owner/operator's fishing time, repair time, decision-making effort, worry, and equity in the business. (Equity is what you own less what you owe.) *$10,655 is the return to labor, management, and equity.*
This is a fair measure of your success from year to year. However, it can also be misleading if you are comparing your business with those of others who have either more or less equity. For ex-

---

[1]   Family members can and in most cases should be paid for their services in the fishing business. Although there are some tax considerations, this is a good way to reward family members and keep their relationship to the fishing enterprise on a businesslike basis.

ample, a friend may have identical costs and returns yet have no debts (unusual but possible!). Interest of $2,200 would not appear under his fixed costs, and his return to labor, management, and equity would be $2,200 greater than yours ($10,655 + $2,200 = $12,855). It appears that your friend is a better fisherman than you, when in fact he may simply have inherited his boat debt-free!

Return to Labor, Management, and Investment

For a true comparative analysis, we must ignore actual interest paid. Therefore, we do not count the $2,200 interest as a cost. We then have the following:

$40,900 – ($21,685 + $8,120 – $2,200 + $440) = $12,855
Gross   Variable   Fixed   Interest Unpaid
returns   costs   costs   paid   labor

The $12,855 is the return to the owner/operator's labor, management, and the $60,000 business investment. If the level of investment changes dramatically during the year, you should calculate the average investment for the year. Suppose that the market value of vessel and gear was $60,000 on January 1, 1974, but in June a $4,000 refrigeration system was installed. The average investment for 1974 would then be as follows:

$$\frac{(6 \text{ months} \times \$60,000) + (6 \text{ months} \times \$64,000)}{12 \text{ months}} = \$62,000$$

The $12,855 is the return to the owner/operator's labor, management, and an average investment of $62,000. This is the most accurate measure of year-to-year success. It is also the most accurate single number by which you can compare your business with other fishing businesses and with nonfishing businesses that require owner/operator labor, management, and investment. Comparing your return to labor, management, and investment with that of similar fishing businesses in your port or region will tell you whether you are a good manager or a poor manager, regardless of changing market conditions, increasing foreign competition, reduced fish stocks, inflation, and so forth. It is a relative measure as well as an absolute measure.

If your return to labor, management, and investment is below that of your fellow fishermen, further analysis is in order. This includes an analysis of the return to labor and management versus the return to investment. It includes an analysis of financial position as well as profit, and an efficiency analysis to identify the weak points of your business. Each of these will be illustrated in this chapter. However, let's jump ahead and assume you have gone through the complete analysis, as indicated above, and have found no way to increase significantly your returns to labor, management, and investment. Now it is time to compare your returns to labor, management, and investment with those being experienced by the local hardware store, service station, sea-food processor, boat yard, and so forth. Review your long-run plans and personal objectives, and decide whether you are willing to accept the lower returns that may result from fishing. Are the nonmonetary benefits from fishing sufficient to offset lower returns and the greater risks involved, or should you sell out and buy a hardware store?

Return to Labor and Management

For the fisherman who has no control over the fishing investment (that is, he leases a boat and gear) and for the fisherman whose only alternatives are working as a crewman, as a wage earner, or in a salaried job, analysis of the return to labor, management, and investment has less value than an analysis of the return to labor and management. Even if you are an owner/operator, it is useful for you to know if your "profit" is due to your skills and decision making or to a productive boat and gear.

How do we separate the return to labor and management from the return to labor, management, and investment? We must use the *opportunity cost* concept. This is explained in some detail in Chapter 5. However, a brief review will be useful.

The opportunity cost of a $1,000 investment is the amount that could be earned on that $1,000 if it were tied up in an equally risky alternate investment. The opportunity cost of $1,000 in your multipurpose bank savings account is the 7 percent interest that could be earned in a savings certificate from a savings and loan bank.

From the $12,855 return to labor, management, and investment derived earlier, we subtract the opportunity cost of the $60,000 investment. What we have left is the actual return to labor and management. What is the opportunity cost of $60,000 invested in a fishing business? Only the investor can answer that. Suppose that you are going to sell out to another, equally competent fisherman, and that you will hold the mortgage. What is the minimum rate of interest you would be willing to accept? Would it be 9, 10, 11, or 12 percent? It will depend upon your estimate of risk in the situation. For our example, let's assume 10 percent.

$12,855 – (10% × $60,000) = $12,855   – $6,000   = $6,855

	Return to labor, management and investment	Opportunity cost of investment	Return to labor and management

The return to labor and management is $6,855. Is this good or bad? It may be bad if you could take a job in the mill or work for the government at $10,500 per year. It may be good if your next best job would pay you only $5,000 per year.

Don't make the mistake of viewing this $6,855 as spendable family income. It is very unlikely to be spendable income because it was derived for purposes of profit analysis. If you are confused about this, review Chapter 5.

Return to Investment

For the fisherman who owns one or more boats and leases them out, an analysis of return to investment is more valuable than an analysis of return to labor, management, and investment. However, if you are an owner/operator, comparing return to investment with return to labor and management can also guide you in improving your business.

The return to investment is separated from the return to labor, management, and investment by subtracting from the latter the opportunity cost of labor and management. Suppose you could sell out and work for the government at $10,500 per year. Your opportunity cost of labor and management would be $10,500. Your actual return to investment would be calculated as follows:

$12,855   – $10,500   = $2,355

	Return to labor, management, and investment	Opportunity cost of labor and management	Return to investment

Is this good or bad? Recall that in calculating the return to labor and management, we decided that we could earn 10 percent on the investment in its next best use. That amounts to $6,000; but we are actually earning $2,355, a 3.9 percent return on investment ($2,355 ÷ $60,000). It is an indication of over-investment relative to the operator's skill and management ability. A decrease in investment or, more likely, an increase in fishing skill and management should improve the situation. The investment needs to be "worked" harder.

Suppose the opportunity cost of your labor and management were only $5,000. The return to labor, management, and investment less this $5,000 would be:

$12,855	− $5,000	=	$7,855
Return to labor, management, and investment	Opportunity cost of labor and management		Return to investment

The $7,855 is greater than the $6,000 opportunity cost of investment referred to earlier. It represents a 13 percent return to investment. It is an indication that some capital improvements in the fishing business could be justified.

Financial Analysis

Five-hundred-page books have been written on financial analysis. We will illustrate only the more important measures of financial condition and success as they apply to a fishing business.

The net worth statement was discussed in Chapter 8. It is a "point-in-time" statement of what you own, what you owe, and what is owed you. It provides the basis for all other financial measures. A completed net worth statement is illustrated in Table 9-2.

E. Z. Fisher has grouped his assets into two categories: current assets and long-term assets. These categories are somewhat arbitrary, though the current asset category will be important in subsequent financial analysis.

Current assets are those that can be converted to cash in a short period of time (usually less than one year) and with little or no expense. The cash in a checking account, in your pocket, or in a passbook savings account is in this category. The accounts receivable

TABLE 9-2.   Net worth statement (balance sheet) for E. Z. Fisher,
F/V *Hi Seas*, January 15, 1974.

Current Assets		Current Liabilities	
Cash on hand	$ 820	Accounts payable at Fleecum Packers	$ 4,150
Account receivable from Fleecum Packers	7,600	Accounts payable at Marine Ways Inc.	2,700
Demand note to Worthy Wally	11,000	Demand note at Center Bank	6,300
Saving certificate	3,000		$ 13,150
Insurance pool rebate	1,400		
	$ 23,820		
Long-Term Assets		Long-Term Liabilities	
F/V *Hi Seas*	$ 94,000	7-year vessel mortgage	$ 58,000
Fishing gear	22,900		
Electronics	8,150	3-year note at Center Bank	1,940
Truck	1,200		$ 59,940
Gear shed	6,500		
	$132,750	Net Worth	$ 83,480
TOTAL	$156,570	TOTAL	$156,570

from Fleecum Packers can be counted as a current asset, provided
that they have a good record of paying these accounts within several
days of demand. A personal demand note to your friend Worthy
Wally can be considered a current asset if Wally is good for the
$11,000 and can pay it within several days of demand.

   Long-term assets are those that normally require more than a
year to convert to cash or that would significantly affect the profitability of your business if converted to cash. These are sometimes
called working assets or fixed assets. All the items listed in this
category could be sold in a few days' time if E. Z. Fisher were to

discount their value. He could probably sell his $94,000 boat quickly if he offered it for $54,000. Because it would take longer to sell the boat for $94,000, and because selling the boat would reduce the profitability of his business, the boat is classed as a long-term asset.

The other side of the net worth statement lists current liabilities, long-term liabilities, and the net worth, which is a liability of the business to E. Z. Fisher. Current liabilities are those that may have to be paid on short notice. Long-term liabilities are those that must be paid at some specific future date or dates.

Net Worth

Assets less liabilities equals net worth. Net worth is what you would have left if you sold out and paid all debts (before taxes on ordinary or capital gains). Net worth is also referred to as owner's equity.

The accumulation of profit and/or an increase in the value of assets increases net worth. Because the market value of boats goes up and down with fishing seasons, fishing regulations, and fish prices, net worth may fluctuate considerably from season to season. This can be offset by adjusting market values once every three to five years rather than once every year. Net worth will then more accurately reflect long-term growth.

Most fishermen would like to see a long-term growth in their net worth. Temporary setbacks may occur from poor years and refinancing. Your net worth is the financial reservoir from which you draw when you are no longer able or interested in working for a living.

Financial Ratios

The net worth statement shows the amount of owner's equity at any time. However, the net worth statement can provide you with much more information concerning your financial condition. You can obtain this information by examining some financial ratios.

The net capital ratio is a good index of the financial security and solvency of the fishing business. Total assets divided by total liabilities gives the net capital ratio. Using E. Z. Fisher's net worth statement, we find that his net capital ratio is $156,570 \div $73,090 = 2.142. The value of this ratio is apparent if we compare Fisher's present net worth statement with a hypothetical statement ten years hence. Assume that in ten years Fisher's net worth has increased from $83,480 to $100,000. Is this an improvement? Not if it has been accomplished with $350,000 in assets and $250,000 in liabilities; this yields a capital ratio of $350,000 ÷ $250,000 = 1.4. In ten years, Fisher has gained $16,520 in net worth but has moved from a position where he had $2.142 in assets for every $1 in liabilities, to one in which he has $1.4 in assets for every $1 in liabilities. His financial security has decreased! This fact would be lost if we looked only at net worth.

The ratio of debt to net worth is another good measure of financial security, financial leverage, and financial flexibility. It is found by dividing total debt by net worth. In Fisher's case, this would be $73,090 ÷ $83,480 = .875. The lower the ratio, the greater the proportion of owner's funds to borrowed funds in the business.

Current assets less current liabilities is a good index of your ability to withstand a short-term financial crisis. In Fisher's case, it is $23,820 in current assets less $13,150 in current liabilities, or $10,670. If Fleecum Packers goes "on the rocks" and defaults on the $7,600 due Fisher, Fisher will probably survive. However, if Fisher is injured halfway through the season and the bank calls its $6,300 note, he may be forced to convert some of his long-term assets and reduce the future profitability of his business.

The current ratio provides similar information and is found by dividing current assets by current liabilities. Fisher's current ratio is $23,820 ÷ $13,150 = 1.81, which is considered low for most businesses. Fisher should consider converting some of his current liabilities to long-term liabilities in order to remove some short-term financial pressure. Too much short-term credit is a common problem among high-risk businesses such as fishing.

There are several other financial ratios that could be calculated, but each is designed for a special purpose and is best worked out with your lender or accountant when the need arises. Decisions based solely upon financial ratios and net worth statements could significantly reduce profitability. Because profit is the best source of increasing net worth and financial security, financial analysis must not be isolated from profit analysis and efficiency analysis.

Efficiency Analysis

Efficiency is the rate at which you convert effort into fish, fuel into knots, costs into revenue, investment into profit, and so forth. It is the ratio of inputs to output in your business. We could there-

TABLE 9-3.   1974 costs and production for F/V *Oltubbe*, owned and
operated by M. T. Nett.

Business investment	$ 86,500
Tons finfish landed	702
Gross returns	$196,560
Days fishing	150
Crew size	5 men
Variable costs (including $5,843 in gear costs)	$ 98,200
Fixed costs	$ 19,100
Total costs (excludes interest)	$117,300
Return to labor, management, and investment	$ 79,260

fore make a ratio from any input (hours fished, hooks used, fathoms
of net, tons of ice, and so on) and output (tons of fish, knots, gross
returns, and so on). Pick the inputs and outputs that you consider
important in your business, and compare them from season to season and with other fishing businesses.

Although profit is the best indicator of success, the ratio of
gross returns to total costs is the best indicator of your efficiency in
attaining profit. For the sake of illustrating this and other ratios,
some cost and production information for M. T. Nett, owner and
operator of a typical 65-foot North Pacific dragger, is presented in
Table 9-3.

Let's first calculate the ratio of gross returns ($196,560) to total
costs ($117,300). This is $196,560 ÷ $117,300 = $1.676. M. T.
Nett was able to obtain $1.68 in gross returns for every dollar he
spent in his business. This is a measure of efficiency, not financial
success. A small business with a gross returns/total cost ratio of
$1.68 may not produce enough profit to support a family, whereas
a very large business with a $1.68 ratio may support several families.
If M. T. Nett can maintain this ratio as his business grows, his profit
will increase. The gross returns/total ratio should be calculated
each year and compared with past years.

Gross returns per dollar invested is another good measure of
economic efficiency. This ratio can be increased by increasing gross
returns for every dollar invested or by allowing the investment to
decline while maintaining gross returns. If investment declines over

time, the profit-earning capacity of the business could be jeopard-
ized in the long run. Therefore, the level of investment as well as
the gross returns per dollar invested must also be measured. For
M. T. Nett this ratio would be $196,560 (gross returns) ÷ $86,500
(investment) = $2.27. M. T. Nett was able to produce $2.27 worth
of fish for every dollar invested. This ratio has a tendency to become smaller as the business grows. If you are increasing your investment — adding refrigeration, buying a larger boat, adding a
larger engine, or whatever — make each new dollar invested work as
hard as the last dollar. Keep this ratio high.

Other good measures of economic efficiency are gross returns
per day fished, gross returns per man-day fished, and gross returns
per dollar of variable cost.

gross returns per day fished: $\dfrac{\$196,560}{150 \text{ days}} = \$1,310$ per day

gross returns per man-day fished: $\dfrac{\$196,560}{150 \text{ days} \times 5 \text{ men}} = \$262$ per man-day

gross returns per dollar of variable cost: $\dfrac{\$196,560}{\$\ 98,200} = \$2$

If you are in a fishery where search is important (tuna, salmon,
hake, menhaden, and so forth), the gross returns per day at sea
would be a more useful measure than gross returns per day fished.

Measures of technical efficiency include tons landed per day
fished, tons landed per man-day fished, variable cost per ton, variable cost per day fished, and gear cost per ton. For M. T. Nett these
would be:

tons per day fished: $\dfrac{702 \text{ tons}}{150 \text{ days}} = 4.68$ tons per day

tons per man-day: $\dfrac{702 \text{ tons}}{150 \text{ days} \times 5 \text{ men}} = .936$ tons per man-day

variable cost per ton: $\dfrac{\$98,200}{702 \text{ tons}} = \$140$ per ton

variable cost per day: $\dfrac{\$98,200}{150 \text{ days}} = \$655$ per day

gear cost per ton: $\dfrac{\$5,843}{702 \text{ tons}} = \$8.32$ per ton

If you are concerned about engine efficiency, then engine costs per day or per ton landed should be monitored. If gear placement and rigging (such as in a trap, pot, long-line, or hook-and-line fishery) is critical, tons per unit of gear is a good measure of efficiency.

None of these measures of efficiency have value in isolation. A season-to-season comparison within your own business should be made. Look for government reports, university publications, and industry journals that report industry measures of efficiency. Rate your business by these industry measures.

10 BUDGETING FOR BUSINESS DECISIONS

Budgeting is one of the most useful management tools available to the decision maker. The budget is a physical and financial plan that projects results before the decision is made and real dollars are committed. There are three types of budgets:

1. Total fishing business budget.
2. Cash-flow budget.
3. Partial budget.

The total fishing business budget is used when major business changes are contemplated. The cash-flow budget is used to plan for financial needs during the year and from year to year. The partial budget is used to project the results of decisions that affect only part of the business.

The purpose of this chapter is to explain and illustrate the partial budget. Chapter 11 is devoted to the total fishing business budget and the cash-flow budget.

Partial Budget Applications

There are many possible changes in the fishing business that will not involve a major reorganization. For example, a change from loran A to loran C will not change engine repair and maintenance, crew-shares, or food costs. Also, installing a refrigeration system will not change electronics repair and maintenance, gear repair and maintenance, or accounting fees. Because these are partial changes in the business, a partial rather than a total fishing business budget can be used.

There are three basic principles in using the partial budget:

1. Include in the budget only those costs and returns that will change if the decision is implemented.

2. Consider non-monetary factors *after* the partial budget is completed.
3. Know the accuracy and reliability of your partial-budget information.

A Partial-Budget Procedure

There are several ways to prepare a partial budget. Regardless of your procedure, it is important to be systematic. Otherwise, one or more important cost or return items are likely to be left out of the calculations.

The following is a partial-budget procedure that minimizes errors. First, assume that you are proceeding with the change in your fishing business. Then work out the answers to the following questions:

1. What are the increased costs? What costs will be added or increased if you proceed with the venture? (Ignore the costs that won't be changed.)
2. What are the decreased costs? What existing costs will be reduced or eliminated if you proceed with the venture?
3. What are the increased receipts? How much will existing income or receipts be increased? What new receipts will there be? (Ignore receipts and income that will not change as a result of the venture.)
4. What are the decreased receipts? What income and receipts will be foregone if you proceed with the venture?

Once these calculations are completed, you need only add the decreased costs (question 2, above) and increased receipts (question 3) and subtract from them the increased costs (question 1) and decreased receipts (question 4). A positive result means that the change would be profitable. A negative result means that the change would not be profitable. Let's illustrate this procedure with a hypothetical example.

Partial-Budget Procedure Illustrated.

Suppose you are the owner and operator of a 54-foot purse-seine salmon boat fishing the North Pacific. The National Marine Fisheries Service offers to charter your boat for gear experiments. The NMFS wishes to have the boat from August 20 to September 15, the middle of the salmon season. They will pay $4,000 for the charter and $1,500 for you to operate the boat, they will provide their own crew, food, and fuel, and they will keep any fish caught. It will be necessary for them to remove some deck gear to make room for their own experimental equipment. You will have to replace the deck gear at a cost of $215.

Your "normal" situation during the August 20 - September 15 period (the time of the potential charter) can be derived from past records and logs, and would include the following:

Gross stock (gross returns)	$9,700
Net repair and maintenance	100
Fuel	600
Food	400
Crew-share (40% of gross stock)	3,880
Vessel repair and maintenance	300
Insurance	1,000
Depreciation	1,000

Using the above information our partial budget looks like this:

1. Increased costs? Taking the charter would require replacing the deck gear at the end of the charter period.   <u>$215</u>
2. Decreased costs? Taking the charter means no net repair or maintenance ($100), no fuel cost ($600), no food cost ($400), and no crew-share ($3,880). Vessel repair and maintenance, insurance, depreciation, and all other costs will not change, so we ignore them. Therefore, decreased costs are: $100 + $600 + $400 + $3,880 =   <u>$4,980</u>
3. Increased receipts? You will gain $4,000 for the boat and $1,500 for your skills and time: $4,000 + $1,500 = <u>$5,500</u>
4. Decreased receipts? Since you will not be able to fish from August 20 through September 15, you forego the $9,700 worth of salmon that would normally have been landed.   <u>$9,700</u>
5. Potential gross benefit (decreased costs plus increased receipts): $4,980 + $5,500 =   <u>$10,480</u>

6. Potential gross debit (increased costs plus decreased receipts): $215 + $9,700 =   <u>$9,915</u>
7. Net benefit: $10,480 - $9,915 =   <u>$565</u>

Keep your budget simple! Don't clutter and confuse your decision with irrelevant information. <u>Consider only the costs and receipts involved in the decision.</u> Categorize them as: (1) increased costs, (2) decreased costs, (3) increased receipts, or (4) decreased receipts.

Making the Decision

A partial budget does not make the decision for you. In our illustration, we came out with a net benefit of $565. This $565 is over and above what could be earned if you salmon-fished. Many fishermen would decide not to take the charter in spite of the $565 benefit. Why?

Contracting for a charter compromises many of the reasons you fish for a living. You would not be independent during the charter period. You would have to tolerate scientists and students on board, and your decks might be cluttered with scientific gear. This may be enough to make you reject the potential $565 net benefit.

If you have little confidence in the accuracy of your budgeting information, you may simply avoid the decision and continue as before. In this case, with good records and logs you could have made a $565 profit decision with confidence.

You may turn down the charter, knowing that the $9,700 gross stock is the average of the past three years, and that next year's may actually be $2,000 to $3,000 higher (However, it may also be that much lower). An "optimist" decision maker will turn down a secure $565 in hopes that next season will be better than predicted.[1]

Other considerations that are likely to influence the decision include: the opportunity to learn something useful from the NMFS experiments, the security of charter income versus a risky fishing income (important if your financial situation is weak), and the chances of getting your crew back if they miss these three weeks of fishing.

Partial-Budget Information

The best single source of information for partial budgeting is your own records, diaries, and logs. The more organized, up-to-date, and complete your records, diaries, and logs, the more efficient and accurate your partial budgeting and therefore your decision making. Other sources of information include data provided by manufacturers and dealers, university and government publications, newspapers, magazines, and, last but not least, other fishermen.

---

[1]   In Chapter 7, a decision maker using the "optimist" criteria was described as a person expecting the best possible result from each decision.

The economic principles illustrated in this book apply to the collection of information as well as to the operation of a fishing boat. It makes little sense to devote days of your time and hundreds of your dollars to collecting information for a $150 decision. However, some decisions may involve a major, long-term change in your business and an elaborate partial budget. In this case, considerable time and expense in collecting information is justified.

You will never have complete, 100-percent accurate information for a partial budget. There are still risks in making decisions. However, with a partial budget and some knowledge of the accuracy of your information, these risks can be taken intelligently. The fisherman who avoids a decision, saying "you can't predict anything in fishing," "you can't trust the information," "things change too fast," "you can't trust anybody," is making a decision to continue as before. This kind of decision maker is likely to be happier working for someone else.

11 BUDGETING FOR FINANCIAL STABILITY

Making major changes in the fishing business without projecting the results on paper is adding risk to an already risky business. The total fishing business budget can help you avoid making financially disastrous decisions when considering major changes.

Being short of cash when an insurance premium or a principal payment must be paid is not only embarrassing but can lead to long-term financial problems. The cash-flow budget is an excellent tool for planning cash inflows and cash outflows in your fishing business.

This chapter explains the cash-flow budget and the total business budget and demonstrates their use in a fishing business.

Cash-Flow Budget

The cash-flow budget is a chart of cash inflows, cash outflows, cash surpluses, and cash shortages for each of twelve or more months or for each of several years. Cash flows are usually prepared on a month-by-month basis and projected for twelve months.

Annual Cash-Flow Budget

Budgeting cash inflows, cash outflows, cash surpluses, and cash shortages allows you to plan for large, one-time cash needs such as loan payments, insurance premiums, and license fees. You may feel that cash flows cannot be predicted for twelve months. This may be true if you have had little experience in the fishery and don't keep logs, diaries, or records. However, you can still prepare an annual cash-flow budget, recognizing that it may not be completely accurate; as you gain experience and progress through the year, you can update your budget and make it more accurate.

The cash-flow budget shows cash inflow, either by item or in total, from such sources as fish sales, the sale of your pickup truck, pay checks, money borrowed, and the sale of fish gear "assembled for sale." Cash outflow should be more detailed: business cash needs that are about the same each month should be separated from the infrequent and large cash needs and non-business cash needs.

Illustration of a Cash-Flow Budget

A cash-flow budget for I. M. Shortt and his boat F/V *DeeFawlt* is illustrated in Table 11-1. It is based upon several years' records, past cash-flow budgets, and some investigation of the future season. Let's study I. M. Shortt's cash-flow budget and see if we can understand how it is organized. Examine Table 11-1 and study the figures on each line as we explain the line headings.

Balance forward is the projected cash on hand — in savings, in the checking account, and available from stocks, accounts receivable, and so forth — carried forward from the previous month. The January balance forward is from December, 1975. The February balance forward is from the bottom line under January, 1976.

Fish sales are the projected gross returns from all fish sold. Receipts are shown for the month when they will be received, regardless of when the fish are delivered.

Sale of capital assets includes the amount realized when you sell an item that has been capitalized and depreciated (its cost was not written off during the year it was purchased). For purposes of profit analysis and tax management, part of the cash received might be treated as ordinary income and the rest as recaptured depreciation and capital gain. For cash flow, none of these distinctions are made — it's all cash inflow. I. M. Shortt plans to sell an old pickup truck in March for $2,200. The entire $2,200 is entered in the cash-flow budget.

Other income includes all other sources of cash inflow except borrowed funds. I. M. Shortt will be working part-time in the cooperative fish plant during the winter months and expects to receive $200 per month. Other possible sources of income include interest on investments or savings, subsidy payments, co-op rebates, tax refunds, gifts, rental income, withdrawals from the capital construction fund, and bonuses.

Total cash inflow is the total cash that will flow into the business during each month. I. M. Shortt may never actually have $12,170 on any one day in March because cash from fish sales, the sale of the pickup, and work at the co-op may come in on different days.

TABLE 11-1. Annual cash-flow budget for I. M. Shortt and F/V *DeeFawlt*, for twelve months beginning January, 1976.

	Jan.	Feb.	March	April	M
Balance Forward	7,050	8,070	470	790	
Fish Sales	6,200	7,800	9,300	14,200	15
Sale of Capital Assets			2,200[1]		
Other Income	200[2]	200[2]	200[2]		
TOTAL CASH INFLOW	13,450	16,070	12,170	14,990	16
Crew Payments	2,230	2,800	3,350	5,100	5
Operating Expenses	1,850	1,850	2,150	3,000	3
Taxes, Fees, and Licenses	450				
Insurance Premiums					
Accounting and Legal		380	80		
Other	100[3]				
TOTAL CASH EXPENSES	4,630	5,030	5,580	8,100	8
Capital Purchases			6,800[6]		
Family Requirements	750	750	1,000	700	
Income Taxes		11,820			
Principle on Long-Term Debt				2,400[9]	
Interest on Long-Term Debt				990	
TOTAL CASH OUTFLOW	5,380	17,600	13,380	12,190	9
Cash Inflow Less Cash Outflow	8,070	−1,530	−1,210	2,800	6
Short-Term Borrowing	—	2,000	2,000	—	
Short-Term Pay-Back	—	—	—	2,000	2
To Balance Forward	8,070	470	790	800	4

1 Sale of truck.
2 Work at fish co-op.
3 Moorage.
4 Boat hauling, etc.
5 Bonus to crew.
6 New truck.
7 New nets.
8 Replace radio.
9 $9,000 outstanding on $12,000 gear loan (5 years @ 11 percent).
10 $50,000 outstanding on $100,000 vessel loan (7 years @ 8 percent).

June	July	Aug.	Sept.	Oct.	Nov.	Dec.
4,780	4,850	10,270	15,310	11,330	17,080	360
15,000	14,800	14,100	13,600	13,600	8,700	8,000
				200[2]	200[2]	200[2]
19,780	19,650	24,370	28,910	25,130	25,980	8,560
5,400	5,330	5,080	4,900	4,900	2,880	3,130
3,500	3,300	3,000	2,900	2,400	1,700	1,500
					370	
5,300						
80				80		120
	100[3]				2,800[4]	300[5]
14,280	8,730	8,080	7,880	7,300	7,750	5,050
			9,000[7]		840[8]	
650	650	980	700	750	750	900
					14,280[10]	
					4,000	
4,930	9,380	9,060	17,580	8,050	27,620	5,950
4,850	10,270	15,310	11,330	17,080	−1,640	2,610
−	−	−	−	−	2,000	−
−	−	−	−	−	−	2,000
4,850	10,270	15,310	11,330	17,080	360	610

Nevertheless, the total amount projected to be available to I. M. Shortt during March is $12,170.

Crew payments are categorized separately because they constitute a major outlay for I. M. Shortt. They could easily be included under operating expenses. Shortt has based his projected crew payments upon his projected fish sales.

Operating expenses include fuel, ice, repairs, food, unloading, and all the expenses associated with fishing. More detail under this category can be justified if your records are detailed enough and if some of these items are of particular concern to you.

Taxes, fees, and licenses constitute a separate category because these cash outflows occur at infrequent intervals during the year. Taxes in this case are on business property only; they are not income taxes. These infrequent expenses usually require more planning than continuing items such as operating expenses.

Insurance premiums include hull, protection and indemnity, health, and accident insurance. If you are incorporated and the business is the beneficiary, include life insurance premiums; otherwise, these appear under *family requirements* (below). Fortunately, insurance premiums are easy to project because they change little from year to year and are due at predictable intervals.

Accounting and legal may be included under operating expenses if your bookkeeping expenses occur monthly. However, I. M. hires a tax accountant in February to handle his tax reporting and an attorney in December to clear some of his crewmen's immigration papers. The amounts involved vary from year to year and require some planning. I. M. has projected accounting and legal fees for February, March, June, September, and December.

Other includes the annual moorage fees, the annual boatyard charges, and the Christmas bonus to the crew (as indicated in Table 11-1, Footnote 5).

Total cash expenses is the monthly sum of crew payments, operating expenses, taxes, fees, licenses, insurance premiums, accounting and legal expenses, and other expenses.

Capital purchases are any purchases that have a potential life of more than one year and that are treated differently than cash expenses for profit analysis and tax purposes. They represent an infrequent and large cash outflow, and are treated separately in the cash-flow budget also. I. M. Shortt projects the purchase of a $6,800 pickup truck in March. Other items that might appear in the capital purchase category are deposits to the capital construction fund, purchases of stocks, bonds, and securities, and deposits to an investment retirement plan (such as, just for a single example, the Keogh plan).

Family requirements are based upon the budgeted family needs for each month of the year. It is a cash draw upon the business, an outflow that depends upon family needs and objectives. I. M. Shortt has projected this to vary during the year. For example, an extra $250 is budgeted for a vacation during March.

Income taxes appear in the month when they are normally paid. Deposits for FICA and withholding taxes on the crew should also appear on this line in the month when deposited.

Principal on long-term debt and *interest on long-term debt* refer to payments upon previous debts (loans, notes, mortgages, and so forth). Any debts expected to be incurred or paid within the budget period are listed, respectively, under *short-term borrowing* and *short-term pay-back* (bottom of Table 11-1).

Total cash outflow is the monthly sum of cash expected to flow out of the business. It represents your month-by-month projected cash needs. I. M. Shortt may never need to expend all of the $14,930 on any one day during June, but during the month of June he expects $14,930 to flow out of his business.

Cash inflow less cash outflow is the net cash situation at the end of each month. I. M. Shortt estimates an $8,070 surplus at the end of January. However, he estimates a $1,530 shortage at the end of February, and he has budgeted a $2,000 loan to carry him through that month. However, another cash shortage will occur in March, and an additional $2,000 loan is budgeted for this.

Short-term borrowing and *short-term pay-back* indicate, respectively, the amounts to be borrowed when cash shortages occur and the amounts to be paid back when cash surpluses occur. I. M. Shortt can use unsecured short-term notes to sustain himself until he has a surplus. During months of surplus he can repay these short-term notes.

Using the Cash-Flow Budget

Without his annual cash-flow budget, I. M. Shortt would have found himself in a difficult financial situation at the end of both February and March. His lender may be willing to loan him $2,000 in February but may question Shortt's financial situation when he returns in March for an additional $2,000. The cash-flow budget gives the lender the "big picture."

Surplus cash can be used to advantage if you know how long it will be available. At a minimum, I. M. Shortt could plan to put

$4,000 in a passbook savings account in May, another $6,000 in July, and plan to draw it out in November when needed. This would provide interest income on the cash that won't be needed for the five months between May and November.

You should not view the cash-flow budget as a financial strait jacket. Rough it out and revise it until you are satisfied. For example, I. M. Shortt's cash situation may look feasible at first glance. He projects a reduction in his debt by $2,400 plus $14,280, and projects several sizable capital purchases. However, Shortt ends the year with a projected cash balance that is $6,440 less than when he started. If 1977 is similar to 1976, he will be forced to borrow for seven instead of three months. This situation could be improved by deferring the purchase of the new truck and/or by reducing the amount of principal payments on the long-term loans. A new cash-flow budget reflecting these changes could be prepared.

If you progress through the year and find each month's experience quite different than what was budgeted, prepare a new six-, eight-, or twelve-month budget reflecting these changes. For example, Shortt may injure himself in February, discontinue working in the co-op, have to hire a skipper for his boat, and have to pay several thousand dollars in hospital bills. A new budget would show the reduced cash inflow, the greater cash outflow, and the adjustments made in capital purchases, family requirements, and so forth, to cope with this situation.

With a detailed cash-flow budget, your lender can decide more easily what credit terms to arrange for you. In fact, most lenders require some form of cash-flow budget. With this budget, you can anticipate large cash requirements and reduce the financial penalties for late payments. You can also schedule your short-term loans so as to minimize outstanding short-term debt and interest.

Total Fishing Business Budget

If you are considering a major change in your business, estimates of future profit, net worth, and cash flow are useful. A common major change is the purchase of a new boat.

Let's assume that our friend I. M. Shortt is considering a second boat for his business. A total business budget for this decision will include a projected profit statement, projected net worth statement, and projected cash-flow budget for each year of the loan.

Projected Profit Statement

Assuming I. M. Shortt obtains a five-year loan on a new boat similar to the old boat F/V *DeeFawlt*, the projected profit statement would look like that illustrated in Table 11-2. The projected profit statement is an annual summary of gross returns, costs, and return to labor and management for the five years of the new loan. Because the F/V *DeeMand* won't be purchased until the middle of the year (1977) and because time is allowed for "shaking down," gross returns and costs for 1977 are increased by less than half of a year's normal gross returns and costs. Increases in gross returns from 1977 to 1980 are due to the increasing productivity of F/V *DeeMand* and an estimated 10 percent annual increase in fish prices. However, gross returns are projected to level off after 1980 due to overexploitation of the fish stocks.

Variable costs are more than doubled when two boats are owned because a skipper for the second boat must be hired. Variable costs are projected to increase through 1981 due to inflation and to increased repairs as each vessel ages.

Fixed costs do not double because such items as accounting, lawyers' fees, and utilities increase only slightly when another boat is added. Fixed costs decline through 1981 due to decreased annual depreciation (a declining-balance method was used).

Opportunity costs of investment are calculated at 12 percent of the average annual investment. The average annual investment is the year-beginning plus year-end value of boats and gear, divided by two.

TABLE 11-2. Five-year projected profit statement for I. M. Shortt, F/V *DeeFawlt*, and F/V *DeeMand* (purchased June 20, 1977).

	1977	1978	1979	1980	1981
Gross Returns	$182,700	258,600	281,800	283,000	283,000
Variable Costs	$131,640	153,500	157,600	172,300	178,000
Fixed Costs	$ 44,800	56,400	52,760	50,810	50,720
Opportunity Costs of Investment	$ 26,500	37,140	33,380	30,100	27,200
Return to Labor and Management	$- 20,240	11,560	38,060	29,790	27,080

TABLE 11-3. Five-year projected net worth statement for January of each
year, I. M. Shortt, F/V *DeeFawlt*, and F/V *DeeMand*.

	1977	1978	1979	1980	1981
Current Assets	$ 610	200	2,000	3,800	5,000
Other Assets					
F/V *DeeFawlt*	$ 90,000	82,800	76,200	70,200	64,500
F/V *DeeMand*	–	182,400	167,800	154,400	142,000
Gear	$ 20,000	61,000	48,800	29,040	31,230
Other	$ 82,720	3,117	21,960	74,800	118,290
Current Liabilities	$ 910	100	1,000	2,000	2,000
Other Liabilities					
F/V *DeeFawlt* Loan	$ 35,720	21,440	7,160	0	0
F/V *DeeMand* Loan	–	$100,000	80,000	60,000	40,000
Gear Loan	$ 6,600	54,200	41,800	30,000	20,000
Net Worth	$150,000	153,777	186,800	240,240	299,020

This figure decreases from 1978 to 1981 due to the declining market
value of boats (8 percent per year) and of gear (20 percent per year).
Average investment for 1977 is less than for other years because
F/V *DeeMand* and its gear were owned for only a half year.

The last row of figures in Table 11-2 is one of the three key
pieces of information needed to make an accurate boat-purchase decision. It shows the return to the owner's labor and management
for each year. This averages out to $17,250 for the five years being
considered. The important question is whether this $17,250 would
be better or worse than the return to labor and management if the
F/V *DeeMand* were not purchased. If the projected difference is
small, the accuracy of the figures in I. M. Shortt's projected profit
statement becomes critical. The information from the projected net
worth and cash-flow statements also becomes more important.

Projected Net Worth Statement

Let's look at a projected net worth statement and see what
would happen to I. M. Shortt's financial situation if he were to buy
F/V *DeeMand*. A five-year net worth statement is illustrated in
Table 11-3. The projected net worth statement in Table 11-3 has
less detail than would be expected in a current net worth statement.
Nevertheless, all the important information is there. Shortt begins
1977 with only F/V *DeeFawlt* and a net worth of $150,000. His
net worth ratio (net worth as a percentage of all liabilities) is a safe
28 percent, and Shortt has $82,720 in savings certificates and stocks.
We saw from the projected profit statement in Table 11-2 that he
could invest this amount plus borrowed funds in a new boat, and
expect to earn a 12 percent return on the investment with an
average annual return to labor and management of $17,250 over
the five years. This 12 percent is better than current earnings on
the savings and stocks, so the investment looks good.

In 1978 a substantial increase in debt is projected ($132,510),
and assets are estimated to increase by $136,187. This increases
the 1978 net worth by $3,677, the gross returns less cash costs from
1977. The net worth ratio drops to a disastrous 114 percent. Shortt's
lenders would have $1.14 in his business for every $1.00 he has in
the business. The lenders would have majority financial control.

The crucial year is 1978. If profit works out as budgeted, the
financial situation improves from 1978 until 1981; by this time,
net worth has nearly doubled and the net worth ratio has dropped
to 21 percent, better than in 1977.

Projected Cash Flow

The projected profit and net worth statements look favorable.
The last budget to estimate is the cash-flow budget for the next five
years. Will there be enough cash inflow to match cash outflow over
the five-year loan period?

If we look at the five-year projected cash flow for I. M. Shortt
(Table 11-4), we see that there will be a cash shortage during 1977,
1978, and 1979. The projected shortage in 1978 is substantial.
Shortt will either have to borrow additional funds to carry him
over these three years, defer some expenses until 1980 or 1981, or
give up buying F/V *DeeMand*.

The figures in the five-year cash flow are derived from the pro-

TABLE 11-4. Five-year projected cash-flow budget for I. M. Shortt,
F/V *DeeFawlt*, and F/V *DeeMand*.

	1977	1978	1979	1980	1981
Balance Forward	$ 83,330	-4,617	-16,827	-227	11,073
Cash Inflow	$183,700	258,600	281,800	283,000	283,000
Cash Expenses	$152,640	191,953	192,670	207,000	211,400
Capital Purchases	$240,000	0	0	0	0
Family Require- ments	$ 9,400	10,300	11,000	12,000	13,000
Taxes	$ 550	7,100	12,000	13,000	12,000
Principle	$ 16,880	46,680	38,960	30,000	30,000
Interest	$ 2,177	15,677	11,570	9,700	5,400
Cash In less Cash Out	-$154,617	-17,727	-1,227	11,073	22,273
Total Borrowings	$150,000	900	1,000	0	0
To Balance Forward	$ -4,617	-16,827	-227	11,073	22,273

jected profit and net worth statements. Family living, depreciation,
and interest are treated differently in each of these statements.
Therefore, the figures are not the same for many items.

A Basic Financial Budget

The best time to develop a five-to-ten-year profit statement, net
worth statement, and cash-flow budget is before they are needed.
Revise them and revise them again. It will take time but they will be
there when you need to evaluate major changes in your business.
Many potentially profitable opportunities won't wait until you develop your budgets. If you have a prepared annual cash-flow budget
and a five-to-ten-year projected profit statement, net worth statement, and cash-flow budget for your existing business, you can modi-
fy them quickly and make accurate and timely decisions when those
great opportunities appear.

12 LEGAL FORMS
OF BUSINESS ORGANIZATION

Most fishing businesses are organized as sole proprietorships. This form of organization has many advantages, but there are also advantages in organizing as a partnership or corporation. In this chapter, we will discuss the advantages and disadvantages of each type of business organization. We will also point out some of the more important considerations in organizing your fishing business.

What is a Legal Form of Business Organization?

To be officially in business, you need only incur some expenses in the pursuit of profit. The clam digger who invests in a shovel, some burlap sacks, and a license, and who then digs, is in business. He need not show minimum expenses or profit. He is in business as a _sole proprietorship_.

As his business grows, he may wish to take on a partner. He can do this by sharing expenses and income with others (an unwritten partnership agreement). He would then be in business as a _partnership_.

Finally, he may decide to incorporate. This, and only this form of business, requires some formal action on the part of state government. The state permits the creation of a new business entity where none existed previously. Our clammer is no longer the business, but part of a new business organization, the _corporation_.

In a sole proprietorship, you and your business are the same. In a partnership, you are only a part of the business — at least in practice. In a corporation, you are only a part of the business — both in practice and legally.

Advantages of a Sole Proprietorship

Simplicity in management. You make all the decisions. You don't need to consult with others. You are the manager and are accountable only to yourself. It is a very uncomplicated arrangement.

Flexibility. Because you are the sole decision maker, it is easier for you to make changes in the fishing business, provided you are financially able to do so. If you decide to quit fishing early in the season, there are no others who can officially object. If you hear that the fish catch is picking up, you can easily revoke your earlier decision and return to fishing. You can take advantage of new opportunities and change your mind quickly.

Personal satisfaction. If the decision turns out well, if you have a profitable season, if you make a good buy on a boat, you take all the credit and satisfaction. Because you are the sole decision maker, you reap all the benefits — both monetary and nonmonetary.

No legal formalities. You can begin a sole proprietorship by merely going into business. You can buy a skiff, some set lines, get any appropriate licenses, and go fishing. There are no other legal requirements.

Termination or modification is easy. If you want to go out of business, you simply quit and/or sell out. If you want to go into fish processing, retailing, or some other business, you go through the normal decision-making process, acquire the facilities, and begin.

Disadvantages of a Sole Proprietorship

Unlimited liability. Your personal and business assets are liable in a sole proprietorship. If you have an injured employee, make a financial mistake, or do anything else for which you are liable, you can lose all your personal and business assets.

Size limited by management. Your fishing business can be only as large and as successful as your management ability will allow. This may not be large enough to satisfy your family objectives. You can expand your management abilities through experience and education, or you can hire management assistance. This will allow further expansion of your business. At some point, however, you may have to give up being a sole proprietor in order for your business to grow.

Size limited by capital. All of your own capital plus what you can borrow may not permit a large enough business to meet your family objectives. There is a limit to every sole proprietorship's borrowing capacity, as illustrated in Chapter 11. This severely limits the success of many fishing businesses.

Discontinuance in case of death. Your sole proprietorship ends with your death. If your son wants to continue the business, the assets may be transferred to him but the cost of doing so can be considerable. He must establish a *new* sole proprietorship.

Advantages of a Partnership

A minimum of legal formalities. A partnership can be as simple as a sole proprietorship or as complicated as a corporation. Most partnerships are verbal and require no legal papers.

Bring together more equity. When two or more persons go into business together, the borrowing power of the business is multiplied by the greater equity contributed by the partners. Where a sole proprietor with a net worth of $20,000 could borrow only another $40,000 to buy a $60,000 boat, two partners with a net worth of $20,000 each could borrow $80,000 and buy a $120,000 boat. This one larger boat might then provide greater profit to each partner than two individually owned $60,000 boats.

Bring together more management ability. Two or more partners in a fishing business may bring together special management capabilities that will produce more profit than would the sum of their individual businesses. For example, Fast Catcher may be a good vessel operator and fisherman but may not know how to handle his money. Sharp Figure may know how to handle money but not a fishing boat. Individually, their chances of success in commercial fishing would be small. As partners, Catcher can concentrate on catching fish, Figure can concentrate on managing the finances, and they will more than double their previous individual profits.

Termination is relatively easy. If the partnership is to be discontinued, one partner need only quit. However, disentangling the finances may be quite complicated.

Disadvantages of a Partnership

Unlimited liability. A partner is liable not only for his own mistakes or misfortunes, but also for those of his partners. A suit against the business leaves both partners' personal and business assets vulnerable. It is possible in some states to establish a limited liability partnership, in which a partner's personal assets are protected from suits against the business. This, of course, makes the partnership more complicated.

Management conflicts. When two or more persons are making decisions about a business, some disagreement and conflict is inevitable. The entire decision-making and management process is more complicated. Each partner must know what the others are thinking and doing. This takes a conscious effort and some time. Some potentially profitable decisions may be foregone in order to avoid conflicts, or may not be made in time because of a partnership communication problem. Establishing clear management responsibilities for each partner when a partnership is formed can help alleviate this disadvantage.

Discontinuance in case of death. The partnership ends when any partner dies. Continuance would be as a sole proprietor, or a new partnership could be formed.

Advantages of Incorporation

Limited liability. By forming another entity, which has its own assets and liabilities, you place a legal barrier between you and your business. Liability from lawsuits is not eliminated, but it is divided. If the corporation is sued, its assets can be taken, some of which is your own equity (represented by stocks). Your personal assets can be taken only if a separate suit is filed against you personally.

Division of ownership. Because the assets of a corporation are represented by stocks, ownership of stocks represents ownership of the corporation. It is convenient to allocate ownership of a $70,000 boat by selling 700 one-hundred dollar shares to corporate members. Each can own any amount from 1 to 699 shares. Otherwise, different parts of the boat would have to be owned by different people. Think of the problems that would result if the engine owner didn't want the engine used on Tuesdays and Sundays, and the winch owner wanted the winch used only on the even-numbered days of the month. Such potential problems are minimized in a corporation.

Continuity of business. The life of the corporation is the life designated at the time of incorporation, unless legal action is taken to terminate it. The corporation transcends the life of any officer or stockholder.

Bring together more equity. As with the partnership, the corporation makes it possible to combine the equity of several persons and to increase the business's borrowing power. Further, a public stock corporation can raise capital by selling stock to the public with future profit as the primary security.

Fringe benefits. Corporations can take advantage of special insurance plans, employee pension plans, annuity plans, and other benefits not generally available to sole proprietorships and partnerships.

Tax benefits. Depending on the size of your fishing business, there *may* be a tax advantage if you are incorporated.

Disadvantages of Incorporation

Complicated management. Most states require officers to be named and a board of directors to meet and submit reports to the appropriate state agency. With multiple owners, officers, and directors, the communication problem increases considerably. However, the responsibility of each of these people should be clearer in a corporation, and management may actually be better than in a partnership. The corporate laws and your own bylaws may force you to be more efficient and effective in management than under a partnership or sole proprietorship, where there are no established rules.

More legal and accounting requirements. In most states, corporations are required to file annual reports and to pay fees. This frequently requires the services of an accountant and an attorney. Also, more paper work and attorney services are required when establishing a corporation.

Considerations in Organizing a Partnership

Partners need not be individual persons. A partnership can be made up of corporations, other partnerships, or associations. The behavior of partners determines whether or not a partnership exists. If one fisherman leases a boat from another — but both share in management, expenses and income — they are a partnership, even if the lease agreement declares that they are not. On the other hand, if one owns some gear and the other owns the boat and makes all the decisions, they may not be partners even though a partnership agreement exists.

In forming a partnership, the following should be considered:

1. Specify how long the partnership is to last. If it is to last for a very short period of time (such as one fishing trip or season), it may actually be a joint venture.
2. Agree upon the type and value of contributions each partner will make to the business. It should be well understood how much labor, capital, and/or management each partner is to contribute. There should be agreement on the value of these contributions and on whether they are being turned over to the partnership, or loaned, or leased.
3. Agree upon procedures for withdrawal from the partnership. How is the money or property to be returned to the contributing partners? How is it to be valued?
4. Agree upon procedures for sharing costs and returns. Generally, the returns from a partnership should be shared in the same proportion as the value of all contributions made by each partner. Without such an agreement, your state statutes may require profits to be divided equally. This may not be equitable in your partnership.
5. Management responsibilities must be worked out before the partnership is organized. Who makes policy decisions? Who makes fishing decisions? Who approves payments? Who commits the partnership to debts?
6. Some plan for arbitrating disputes should be prepared. Frequently, a nonpartner is needed to resolve disputes and to prevent the expensive breakup of a partnership.

There are two types of partners: general and limited. There is no protection for the general partner, but he can participate fully in the partnership. As mentioned earlier, a limited partnership helps reduce the partner's liability: it is no greater than his investment in the partnership. However, he does not participate in the management of the business. Independent actions of the limited partner usually cannot dissolve the partnership.

Partnership profits are divided among the partners and are taxed as income to those individuals (whether or not they actually receive the money). The partnership does file an informational return to the Internal Revenue Service, but pays no income taxes.

Considerations in Organizing a Corporation

The stockholders of a corporation are the owners. They may hold any quantity of stock, with each share representing one vote (unless the corporation issues nonvoting stock). The stockholders elect a board of directors who in turn hire a manager and other officers. The board of directors is responsible for policy and major decisions. The manager and his staff run the business. In a small, family-owned fishing business corporation, the stockholders, directors, and officers may be the same persons.

The corporation is made possible by state statutes. Therefore, doing business in another state requires clearance with that state's incorporation agency, as well as with the home state's agency.

To avoid errors that may be costly in the future, the services of incorporation specialists (attorneys, accountants, and so forth) are advisable. You will need their services when applying for your charter and when preparing the bylaws and articles of incorporation.

In assembling a corporation, the following should receive special consideration:

1. Decisions must be made regarding the type, amount, and value of stock.
2. Regulations regarding the issuance, transfer, and repurchase of stock should be developed. If stock is to be held by family members and close associates only, provisions should be made for the corporation or existing stockholders to have first options on any stock that becomes available in the future.
3. Duties, obligations, and salaries for directors and employees should be agreed upon before incorporation.
4. Management responsibilities, accounting procedures, and procedures for reporting to stockholders should also be agreed upon before incorporation.

A corporation may elect (if it qualifies) to be taxed as a partnership. This is frequently advantageous for small fishing business corporations. However, there may be tax advantages in being taxed as a corporation.

Though individual tax rates increase in many graduated steps, there are only two tax rates for corporations. Depending upon the salaries paid the directors and officers, there is a small range of profit where the tax paid per dollar of profit is less for a corporation than for an individual. Estimating future taxable income for your business and calculating taxes on the basis of a corporation *and* as a nontaxed corporation (in which all income is passed directly to stockholders) will help you decide which is more advantageous.

The Fishery Co-operative

The co-operative is another form of business organization available to the fisherman. Most co-operatives are organized to provide marketing, selling, or processing services to members. However, a co-operative may also be in the business of fishing.

Fishermen own a co-operative by owning capital stock or by paying membership fees. Each member has one vote, as opposed to a corporation, where a stockholder has as many votes as shares of stock. Therefore, co-operative ownership and control are vested equally in each member.

The co-operative can be in the form of a partnership or a corporation, depending upon state statutes and the needs of the co-operative. If it is organized as a partnership, the advantages, disadvantages, and considerations involved in a partnership apply. If it is organized as a corporation, the advantages, disadvantages, and considerations involved in incorporation apply.

However, the co-operative has some unique features. The Federal Fishery Co-operative Marketing Act of 1934 allows a co-operative to avoid some of the restrictions in interstate and foreign commerce that apply to corporations. The co-operative can also enter into a variety of agreements with suppliers and customers, some of which are illegal for corporations. To qualify for these special privileges, the co-op must do more business with members than with nonmembers, allow only one vote per member, and pay less than 8-percent dividends on capital stock.

Good management is as important in a fishing co-operative as in any other form of business, a consideration that is often overlooked. Also, because the fishermen are the owners and members, it takes more time and effort to keep them informed and happy. Last, the fishermen must have a certain commitment to the co-op in order to make it successful.

PART THREE

APPLICATIONS

13   SELECTING A FISHERY

How do you decide whether you will fish for cod, shrimp, sole, lobster, mullett, crab, abalone, or pollock? For many fishermen, this decision is made by following in their fathers' footsteps. For others, it's the traditional fishery in their port. You may have happened upon the "perfect" boat and wound up in a fishery for which it was designed and rigged. You may also have heard about the thousands of dollars the shrimpers, lobster trawlers, or king crab boats "made" last season, and you may decide you're going to get some of that action.

Deciding whether or not to fish for a living is often equally haphazard. It often begins with a nonfisherman acquiring that "perfect" bargain boat and then wondering how he will pay for it.

In this chapter, we will apply some economic concepts and management principles to these infrequent but important decisions.

To Fish or Not to Fish for a Living

In deciding whether or not to fish for a living, you should compare the potential returns to labor, management, and investment from fishing with those from your present occupation. You should also review your personal and family objectives, and determine whether fishing will help you realize those objectives.

Let's first consider the potential returns to labor, management, and investment. If you are a schoolteacher with a savings account at the local bank, your return to labor and management is your salary, say $12,500 per nine months, and your return to investment is the 5, 6, or 7 percent in interest you are earning in your savings account.

If you own and operate a small business such as a restaurant, garage, or farm, you calculate returns to labor, management, and investment as illustrated in Chapter 9. Suppose that your restaurant is earning $35,000 over variable and fixed costs and that you have no debt. Your return to labor, management, and investment is $35,000. If you sell out, reinvest all your money in a fishing business, and spend as much time fishing as operating the restaurant, the $35,000 will be directly comparable: can you generate $35,000 over and above your variable and fixed costs in a fishing business?

Although the cost and earnings studies listed in the bibliography can help you answer this question, you should spend some time analyzing your own information. We will illustrate how this can be done in the next section of this chapter.

In addition to or in spite of the potential profit in fishing, many enter the business because of independence, challenge, personal satisfaction, pride, tradition, and pleasure. For many, these benefits help to rationalize an unprofitable business. It is sometimes easier to find non-monetary benefits in the fishing business than to improve its profitability.

Deciding on a Fishery

Assume that you have decided to become a commercial fisherman and are looking for a fishery. You can choose on the basis of the following:
1. Potential profitability.
2. Potential risk.
3. Experience required.
4. Investment required.
5. Time required.
6. Complementarity with other fisheries or occupations.

Once you have purchased a boat, you must also consider its capabilities.

You should investigate these factors and measure them against your objectives before you choose a fishery. Because this decision is rarely made more than four or five times during your career, you can easily justify investing several weeks and several thousand dollars in choosing a fishery.

We will follow the decision-making process discussed earlier, look at each of the six factors identified above, and then illustrate the overall process of selecting a fishery.

TABLE 13-1. Return to equity (net worth) for twelve fisheries, data covering from one to seven years.

FISHERY	1960	1961	1962	1963	1964	1965	1966	1967	1968	1969	1970
alibut [1]							$6,534	4,067	4,877		
una (seine) [1]			$29,719	10,058	14,370	26,775	21,243			72,235	
una (bait boat) [1]			$35,772	19,410	225	2,876					
allops [1]	-$1,515							957	6,378		
hrimp (Atlantic) [1]							$3,756				
rimp (Gulf) [1]					$2,721	5,162	5,946	8,111	6,600	3,240	1,480
enhaden [1]					$5,370	13,311					
ng Crab [1]								$15,498	8,459	2,041	
ungeness Crab [1]								$3,956	6,816		
orthern Lobster [1]								$4,110			
oundfish (Pacific) [2]									$2,360	4,442	
oundfish (Canadian Atlantic) [3]							$2,642		-4,541		

From the United States Department of Commerce, National Oceanic and Atmospheric Administration, Current Fisheries Statistics, Basic Economic Indicators, number 5934 and numbers 6127 through 6133.

From author's unpublished research.

From Proskie, Primary Industries Study No. 1, Economics Branch, Fisheries Service, Department of Fisheries and Forestry, Ottawa, Canada.

Identifying a Profitable Fishery

A profitable fishery can be identified by:
1. Studying the profit history of various fisheries.
2. Estimating the future availability of fishery stocks in those fisheries with profitable histories.
3. Projecting future regulations.
4. Estimating future costs of participating in the fishery.
5. Predicting the future demand or the price of the fishery product.

Some fisheries have been consistently more profitable than others, even though some fishermen in profitable fisheries fail to make any profit. These include the West Coast salmon fishery, the tuna fishery, the shrimp fishery, and parts of the lobster fishery. Other fisheries have been highly profitable in the development stages, only to succumb to overexploitation and regulation. The Alaska king crab fishery is a classic illustration of this, though it is again yielding substantial profits, in part because of rapidly rising crab prices.

Historical profit information is available from several sources. The most comprehensive source is the National Marine Fisheries Service's series entitled *Basic Economic Indicators*.[1] Profitability data from these reports are summarized in Table 13-1. This table

[1] See bibliography.

illustrates the return to equity (or net worth) for twelve fisheries for different years. As shown in the table, the tuna seine and king crab fisheries have been quite profitable in the past. On the other hand, the scallop and Atlantic groundfish fishery have yielded very little profit.

Once you have identified several fisheries with traditionally high profits, you should then study the potential continued availability of the fishery resource. This depends upon domestic exploitation, foreign exploitation, artificial enhancement, and natural variability in populations. The Atlantic cod, yellowtail flounder, and Pacific ocean perch fishery are examples of fisheries that remain unprofitable due to foreign overexploitation.

Profitability in the Pacific Northwest salmon fishery has been sustained in part by multimillion-dollar hatchery programs. This is in spite of greater exploitation. The Pacific Northwest dungeness crab fishery and Chesapeake Bay oyster fishery are subject to widely varying populations. These population variations are apparently due to such natural phenomena as disease, available nutrients, water temperatures, and salinity. There has been some artificial planting (enhancement) of Chesapeake Bay oysters but no enhancement in the dungeness crab fishery. Information from state fishery agency biologists and NMFS biologists can help you estimate future fish availability.

State and federal regulations should also be considered in choosing a fishery. The federal government has yielded most of the regulatory rights to each state. This means the state can regulate fish landed within the particular state, fishermen licensed by the state, vessels licensed by the state, and any living marine resource within adjoining waters.

In addition to state regulations, there are multilateral commissions that regulate one or more fisheries regardless of geography. Interviews with the responsible decision makers in the appropriate regulatory agency are probably the best way to predict regulations for a particular fishery. More regulations generally mean higher fishing costs. Higher fishing costs can also result from reduced fish stocks, higher prices of gear, fuel, supplies, and so on, and more sophisticated equipment.

The prices of gear, fuel, ice, and equipment have been rising so consistently in the past five years that continued increases are quite predictable. Following the wholesale price indexes published in the major newspapers can be helpful. However, because price increases will apply to all fisheries, it is not critical to choose among different fisheries on this basis.

More critical is the potential new technology that may force you to substitute machinery for labor, a larger for a smaller boat, or electronics for intuition. Interviews with experienced fishermen, NMFS gear specialists, and university gear specialists will help you estimate gear costs and other increased fishing costs.

Choosing a fishery with historical profits, lots of available fish, few regulations, and predictable costs will be of little use if nobody wants to buy your product. In the United States, we have large stocks of easy-to-harvest fish that continue to be ignored because the American consumer does not want them. Discuss each fishery with your fish buyer or processor and his broker and wholesaler. Information on future domestic and foreign markets can also be obtained from the NMFS.

A Risky or a Safe Fishery?

Your personal aversion to risk, your financial situation, and your financial responsibilities will determine whether you choose a risky fishery or a safe fishery. In Chapter 7, risk was defined as a situation where the outcome will vary but can be predicted with some probability. Some fisheries are more variable than others and are therefore more risky. The more risky fishery may produce consistently higher or consistently lower profits. Most fishermen are willing to enter a more risky fishery only if there is a probability of high profits.

There are several sources of risk in a fishery. Profit may vary from year to year. This could be due to variable fish populations, variable fishing costs, different levels of exploitation, and/or varying product prices. Another source of risk is the degree of exposure to natural elements. The small tuna troller venturing 200 or 300 miles from port in search of tuna is exposed to more physical risk than the gill-net fishermen in the protected waters of Puget Sound or the Columbia River.

The fisherman with little equity in his business, no dependents, and many years of fishing ahead of him will be willing to enter a much more risky fishery than the fisherman who has 60 percent equity, is making large principal payments, is supporting several dependents, and has seen his fiftieth birthday.

Profit can vary from season to season and from fisherman to

fisherman. The seasonal profit variation is due to seasonal variation in landings, costs, and prices. Variation in profit from one fisherman to another is due to fishing skills and management ability.

We can illustrate the risk for different fisheries by examining the variability of landings and prices over time. The standard deviation is a statistic that measures this variability. The standard deviation tells you that two thirds of the time, the year's landings (or prices) will be no more than the average plus or minus the standard deviation value. The larger the standard deviation, the greater the variability from year to year.

The standard deviations of landings for eleven fisheries over an eleven-year period are shown in Table 13-2. In terms of landings, the king crab fishery is far more risky than the blue crab fishery. The king crab standard deviation of landings over eleven years is 83.4 percent of the average landings, whereas for blue crab it is only 3.9 percent of the average landings. Two thirds of the time king crab landings varied as much as 83.4 percent from the average landings, whereas two thirds of the time blue crab landings varied as much as 3.9 percent from the average. If you are not a risk taker, the fisheries at the bottom of Table 13-2 are not for you!

The standard deviations of ex-vessel prices for eleven fisheries over a ten-year period are shown in Table 13-3. Oysters have the smallest standard deviation as a percentage of average prices. Again, king crab shows the most variability with a standard deviation of 53.9 percent of average prices. This can be misleading, however, for the variability in king crab prices is mostly an upward trend from 1962 to 1971. The same upward trend in prices also applies to the scallop, lobster, tuna, and dungeness crab fisheries. In the last several years, there has been a steady upward trend in prices for nearly all fisheries.

Is Experience Important?

Most fishermen will agree that an important factor for success in any fishery is the skipper's skills and talents. Nearly all fishing skills are learned through fishing. There is no really effective way to learn how to set a net, how to haul a pot, how to find fishing water, and so forth, on dry land. Experience is the best teacher in this case. Therefore, experience is important.

In selecting a fishery, you should capitalize upon your experience, if you have it. If you don't have the experience, you should

TABLE 13-2. Eleven-year, per-boat average landings and standard deviations for eleven fisheries.

Fishery	Years of Observation	Average Landings	Standard Deviation in Average Landings	Standard Deviation as a Percentage of Average Landings
Blue Crab	1959-1969	20,081	784	3.9
Lobster	1956-1966	3,595	242	6.7
Shrimp	1959-1969	24,834	3,137	12.6
Tuna	1959-1969	163,999	27,120	16.5
Oyster	1956-1966	7,049	1,171	16.6
Salmon	1959-1969	18,473	3,430	18.5
Dungeness Crab	1959-1969	42,937	10,930	25.4
Halibut	1959-1969	127,576	35,280	27.6
Clam	1956-1966	81,354	25,600	31.4
Scallop	1959-1969	110,301	38,900	35.2
King Crab	1959-1969	223,500	186,550	83.4

Source: USDC, NOAA, Current Fisheries Statistics, Basic Economic Indicators, number 5934 and numbers 6127 through 6133.

TABLE 13-3. Ten-year average prices and standard deviations for eleven fisheries.

Fishery	Years of Observation	Average Prices	Standard Deviation in Average Prices	Standard Deviation as a Percentage of Average Prices
Oyster	1959-1968	.50	.04	8.0
Shrimp	1962-1971	.59	.07	11.9
Salmon	1962-1971	.20	.03	15.0
Dungeness Crab	1962-1971	.19	.03	15.8
Clams	1959-1968	.25	.04	16.0
Tuna	1962-1971	.15	.03	20.0
Halibut	1962-1971	.28	.06	21.4
Lobster	1959-1968	.64	.14	21.9
Blue Crab	1962-1971	.08	.02	25.0
Scallops	1962-1971	.81	.35	43.2
King Crab	1962-1971	.17	.09	53.9

Source: USDC, NOAA, Current Fisheries Statistics, Basic Economic Indicators, number 5934 and numbers 6127 through 6133.

consider a fishery where experience can be gained as an apprentice — where the financial and physical risk to you will be lower. You can also select a fishery where experience is less important, the technology is less complex, and fewer skills are required.

Fortunately, the skills learned from one fishery are readily transferred to many others. The skills required in trolling for salmon are similar to those required in trolling for tuna. If you are skillful with a herring seine, you are likely to be skillful with a salmon seine. Fishermen who are successful at drum trawling have had little trouble succeeding at twin trawling.

At the minimum, you should talk with several experienced fishermen in the fishery of your choice and fish with them for a trip, a month, or even a year. Use every opportunity to gain experience on deck, in the galley, in the wheelhouse, in the engine room, and on the dock. Observe, ask questions, and take notes. Experience is an investment in future profitability.

Different Fisheries Require Different Investments

Some fisheries require an investment of at least a million dollars, and in others you may need only a five-dollar shovel and ten dollars worth of burlap sacks. Your ability to raise and sustain the capital required is important in selecting a fishery.

Table 13-4 shows the approximate amounts of capital required for sixteen different fisheries. These figures are only approximate because new boats and new equipment may require twice as much capital as used boats and equipment. Also, for some fisheries there is a considerable range of vessel sizes and equipment requirements within which you can succeed. For example, a 60-foot, 30-year-old New England dragger with only a fish finder and loran on board may be as successful as a new 90-foot dragger with six different nets, two radars, and two fish finders.

Table 13-4 also shows the range of annual operating capital required for sixteen fisheries. Operating-capital requirements are often overlooked in selecting a fishery, but they can be quite important. In some fisheries, such as the Alaska salmon fishery, it is common for the fish plant to advance nearly all operating capital. In others, you must finance your own fuel costs, ice costs, repair costs, license fees, and crew-share until you receive payment for your fish. This may be several days or several months later. You must be able to borrow or generate enough operating capital within your business to meet these short-term obligations.

Investment capital comes from your own equity and from long-term borrowing. Most lending institutions require you to have at least 40 percent equity — that is, they will provide up to 60 percent of your total investment needs. However, fishing businesses with 60 percent owner's equity and 40 percent borrowed capital are more common than those with 40 percent owner's equity and 60 percent borrowed capital. Using this as a standard, if you have $20,000 in equity and no experience, the maximum capital you can put together would be $33,000 (60 percent × $33,000 = $19,800, and 40 percent × $33,000 = $13,200). This would eliminate the shrimp, groundfish, king crab, lobster trawl, salmon seine, halibut, and dungeness crab fisheries from your list of choices. However, if you have $60,000 in equity, you could put together $100,000 in capital and enter nearly any fishery.

Cash-flow budgets, projected profit statements, and projected net worth statements (as illustrated in Chapter 11) are the appropriate decision-making tools in selecting a fishery. Such projections will point out which fisheries are within your financial capabilities and which will stretch your capital position to the point of bankruptcy.

TABLE 13-4. Investment capital and operating capital required for sixteen fisheries.

Fishery	Total Investment Capital	Annual Operating Capital
Salmon (troll)	$ 8,000- 50,000	$ 2,000- 8,000
Tuna (troll)	20,000-100,000	15,000- 30,000
Shrimp (Pacific)	90,000-300,000	35,000- 65,000
Groundfish (Pacific)	90,000-300,000	40,000- 70,000
Crab (King)	150,000-550,000	40,000- 90,000
Groundfish (New England)	100,000-350,000	50,000- 95,000
Lobster (trawl)	150,000-300,000	90,000-160,000
Salmon (seine)	90,000-140,000	15,000- 30,000
Halibut (long line)	60,000-150,000	60,000-100,000
Clam (mechanical tong)	30,000- 60,000	10,000- 15,000
Salmon (gill net)	10,000- 50,000	3,000- 8,000
Oysters (dredge)	30,000- 70,000	15,000- 25,000
Shrimp (Gulf)	100,000-350,000	20,000- 50,000
Crab (Blue)	3,000- 30,000	2,000- 10,000
Mullet	3,000- 10,000	2,000- 10,000
Crab (Dungeness)	60,000-180,000	8,000- 20,000

A Full-time or a Part-time Fishery?

For a very large number of fishermen, the most important factor in selecting a fishery is whether or not it will complement another occupation. If fishing is not to be a full-time occupation for you, the fishery you choose is likely to be seasonal and require a moderate investment.

If you are planning to fish during the part of the year when your other job doesn't demand your presence, then the fishery season must match your "time off." For example, if you are a schoolteacher, a logger, or a fuel oil truck driver, a summer fishery that doesn't require too much investment is the logical choice.

You may work part-time by fishing one year and not the next, or by fishing only a few months each year. If you fish some years and miss some years, a more logical choice may be to participate in a year-round fishery or fisheries and lease your vessel when you are not fishing. Selecting a potentially profitable fishery, evaluating its risk, and preparing projected profit, net worth, and cash-flow statements is as important for "part-time" fishermen as it is for full-time fishermen.

TABLE 13-5. Hypothetical total fishing business budget for black cod fishing with traps (information from other fishermen, buyers, and research).

Variable Costs	
Vessel repairs (same as with dragging)	$ 8,000
Gear repairs (600 pots, 200 replaced per year at $70 each)	14,000
Fuel (similar to crab boats)	800
Food	700
Bait	1,100
Crew-share	14,175
	$38,775
Fixed Costs (same as with dragging)	8,000
TOTAL COSTS	$46,775
Gross Returns	
675,000 pounds at $.07 per lb.	$47,250
RETURN TO LABOR, MANAGEMENT, AND INVESTMENT	$475

Some Fisheries are Complementary

It may be profitable to participate in two, three, or more fisheries each year. A certain type of boat may be adapted for several fisheries at little cost. Changes in fish prices, fishing costs, and regulations or the ending of one season and the beginning of another can make it desirable or even necessary to convert to a different fishery.

A classic example of seasonal complementarity has resulted in the creation of the West Coast combination boat. Usually 35 to 80 feet long, this boat trolls for salmon from June to August. In August albacore tuna appear, and after a one- or two-day changeover, tuna is fished until October or November. By December, these boats have their crab blocks mounted and are setting pots for dungeness crab; this is done through March and April. After the usual month lay-up for repairs, the cycle begins again.

This type of complementarity helps reduce risk and increases returns to investment. As indicated in Chapter 7, the advantages of participating in several fisheries must be weighed against the greater variable and fixed costs of a multipurpose boat.

Pot M. Cod Chooses a Fishery

Pot M. Cod has been dragging for flounder and rockfish for several years, but isn't very satisfied with his returns to labor, management, and investment. He has been observing some experimental fishing for black cod with traps and the rapidly developing shrimp fishery. In deciding whether to attempt a black cod fishery and/or shrimp fishing, Pot M. Cod will consider:

1. Potential profit.
2. Risk.
3. Investment.
4. Complementarity.

Careful budgeting will give Cod a good estimate of profitability. Because Cod has no experience and hence no records on which to base his budgets, he must do some research. Cost estimates must be obtained from other fishermen, university studies, and National Marine Fisheries Service research. A fairly complete budget will be used because changing fisheries represents a major change in the business. Table 13-5 illustrates a total fishing business budget for black cod

TABLE 13-6. Hypothetical total fishing business budget for shrimp fishing
(information from other shrimpers, buyers, and research).

Variable Costs

Vessel repairs (same as with dragging)	$ 8,000
Gear repairs (figure obtained from other fishermen)	4,500
Fuel	2,700
Food	1,600
Crew-share	14,700
	$31,500
Fixed Costs (same as with dragging)	8,000
TOTAL COSTS	$39,500

Gross Returns

210,000 pounds at $.28 per lb.	$58,800
RETURN TO LABOR, MANAGEMENT, AND INVESTMENT	$19,300

fishing, and Table 13-6 illustrates a total fishing business budget for
shrimp fishing.

Pot M. Cod's return to labor, management, and investment has
averaged $14,000 for each of the last three years. The black cod pot
fishery with a budgeted $475 return to labor, management, and in-
vestment doesn't appear to be a very good alternative, especially because more investment in the form of pots would be required. The
shrimp fishery, with a budgeted $19,300 return to labor, manage-
ment, and investment, deserves further examination.

How risky is shrimp fishing compared with dragging for flounder
and rockfish? The physical risk would be about the same because
distance and time would be similar and the boat would be rigged and
fished in a similar manner. Landings and prices have been more
stable in the groundfish fishery than in the shrimp fishery. However,
there has been a steady upward trend in shrimp prices. Gross returns
in the shrimp fishery are therefore more likely to keep pace with rising costs.

Because there is some difference in investment between the
groundfish fishery and the shrimp fishery, it is useful to prepare a
projected net worth statement for each. These two net worth statements are illustrated in Table 13-7. Current assets for both shrimp

TABLE 13-7. Projected net worth for two fisheries (hypothetical data).

	Groundfish Fishery	Shrimp Fishery
Current Assets	$ 2,000	$ 2,000
Other Assets		
Fishing Boat	82,000	82,000
Gear	12,000	22,000
Electronics	4,500	6,200
Other	5,850	0
Current Liabilities	1,200	1,200
Other Liabilities		
Fishing Boat Loan	31,000	31,000
Gear Loan	0	5,850
Net Worth	$74,150	$74,150
CURRENT RATIO	1.66	1.66
NET WORTH RATIO	.43	.51

fishing and the present fishery are the same ($2,000). Also, the market value of the boat is the same for both alternatives. Shrimp fishing will require $10,000 worth of new gear and $1,700 worth of new electronics. This is financed by using $5,850 from savings and borrowing $5,850. This increases Cod's liabilities and assets by $5,850. Net worth is the same for both fisheries. The current ratio is also the same for both fisheries, but the net worth ratio is slightly worse. Liabilities as a percent of net worth are .43 for groundfish and .51 for shrimp fishing. The total impact of Cod's financial condition the first year of conversion is quite small. However, the relatively higher profit from shrimp will improve the year-end net worth.

Cod's boat is presently rigged for fishing groundfish. Altering it for shrimp fishing requires no structural changes and few deck changes. The boat can also be readily converted back for fishing groundfish.

The evidence points to a shrimp fishery for Pot M. Cod. All he must do now is make the decision.

14 BOAT SIZE
AND OWNERSHIP

Should I lease or buy my fishing boat? How should I lease or buy? What size boat should I lease or buy? These decisions are not made frequently, but they are very important. In fact, we can use nearly every fishing business management concept and tool in making these decisions.

In this chapter we will discuss advantages and disadvantages of leasing versus buying, and compare various lease and purchase arrangements. We will also discuss economies and diseconomies of size and the boat size decision.

Ownership versus Leasing

The desire to own a boat is strong. There are many noneconomic as well as economic benefits to ownership. However, there are many advantages to leasing and you should consider them before you commit yourself to ownership.

We will first analyze the advantages and disadvantages of both ownership and leasing, and then we will discuss various lease arrangements.

Advantages of Ownership

A certain amount of psychological if not financial security comes with boat ownership. You know that the boat is available if and when you want to fish. There is no concern over finding a boat for next season. Your decision-making efforts can be devoted to other aspects of the fishing business.

The financial security aspect of boat ownership can also be significant. Boat ownership is one of the easiest and most common ways to build equity in your business. It is frequently good business to invest in your own business rather than in the stock market, the bank, and so forth. In owning your own boat, you can use profit to reduce debt and therefore increase your net worth. In fact, most fishermen are forced to do this by making the annual principal payments that are required by lenders.

Another advantage is the familiarity you gain with one boat. In owning and fishing with the same boat year after year, you develop those critical skills that make the boat and gear an extension of your own mind and hands. You become more skillful in fishing and can more accurately determine what improvements in boat and gear will add to profit.

Because you own the boat, you are more likely to invest in improvements that will increase the market value of the boat or will add to profitability over an extended period of time. Boat ownership provides some assurance that you will gain from such capital improvements.

Many fishermen choose to own their boat in anticipation of windfall gains that may come from resale at inflated prices. Although it is true that fishing boat market values have increased rapidly in some fisheries, speculating in the boat market is not without risks.

Other things being equal, most fishermen would choose to own their own boat because of the prestige that goes with ownership. With few exceptions, commercial fishing boats become very personal pieces of property that instill great pride in the owner. The continued existence of small custom-boat builders in spite of the economic advantages of "assembly-line" boats attests well to the importance of pride in ownership.

Disadvantages of Ownership

Perhaps the greatest disadvantages of owning a boat is the required commitment of your own capital. Unless you are one of the fortunate few who inherited a boat debt-free, you will have allocated a large part of your scarce capital to boat ownership. This considerably reduces the other options available to you. In fact, many fishermen find that they have so much of their capital committed to boat ownership that they are unable to purchase needed gear, fuel, electronics, and so forth.

Committing a large portion of your capital to boat ownership also reduces your ability to weather a short-term financial crisis. In Chapter 11, we illustrated the dangers of a low net-worth ratio. This need not be a problem if sufficient capital is available, but it is one of the disadvantages of ownership.

In addition to reducing your financial flexibility, ownership can reduce your operational flexibility. Being committed to one vessel restricts you to one, two, or possibly three fisheries. As the fortunes of these and other fisheries rise and fall, the cost of converting your boat or of selling your old boat and buying a new one may deter you from changing fisheries. One solution is to invest in a multipurpose boat. But although a multipurpose boat may be flexible, it may not be as efficient in each of the different fisheries as a specialized boat. In owning a boat, you either give up the flexibility and specialize, regardless of the fortunes of that fishery, or you buy a multipurpose boat and yield some efficiency to your fellow fishermen.

Flexibility also applies to ease of entry and exit from the industry. Ownership tends to commit you to fishing. In Chapter 5, we illustrated the consequences of boat ownership (fixed costs) when fishing doesn't produce enough gross returns to cover the variable costs. Ownership brings with it fixed costs and other commitments, such as payment of principal, whether fishing is good or poor.

Although fishing boats provide security for loans, they are not the most desirable collateral from the lender's standpoint. Boats have a tendency to sink or burn at inopportune moments, and their value depends upon uncertain industry conditions as well as upon your management. Other property, such as land, permanent buildings, or blue-chip stocks, provides more secure collateral. If your capital is tied up in a boat, you are unable to provide such collateral.

Finally, the pride that comes with ownership has the unfortunate tendency to color your fishing business decisions. Some choices may be made that add little to profitability but will make your boat the most admired one in port. If this is an important objective, there is no conflict with ownership. However, it is important to recognize such decisions for what they are — an investment in pride, not necessarily in profit.

Advantages of Leasing

There are many types of lease or rental arrangements. They range from the gross share lease to joint ownership and profit sharing. Though the share lease is nearly the opposite of ownership, joint ownership and profit sharing are nearly identical. Therefore, our discussion of leasing advantages will be quite general.

Leasing makes it possible for you to be a skipper without a large commitment of capital. You may have insufficient capital to acquire ownership, or you may be unwilling to invest your capital in a fishing boat until you have gained some experience, demonstrated the profitability of fishing, proved other investments to be less profitable, or made a final decision to become a fisherman. As indicated earlier, leasing provides more flexibility from season to season. As the profitabilities of different fisheries rise and fall, different boats can be rented or leased to take advantage of the more profitable fisheries.

Leasing also makes it possible for you to obtain larger, more efficient boats and better gear at an earlier stage in your fishing career. Many fishermen who insist upon ownership don't acquire an optimum boat-and-gear combination until they are near retirement age. With a lease arrangement, boat size is limited primarily by your management ability and fishing skill, secondarily by capital availability.

Disadvantages of Leasing

Perhaps the greatest disadvantage of leasing is the lack of security. Your means of obtaining a living are not completely under your control. The boat owner has the option of leasing his boat to another fisherman, or of fishing it himself.

Another problem arises if you wish to make improvements on the boat. It may be difficult to convince the owner of the need for capital improvements if you are the benefactor. On the other hand, if you finance the capital improvements, you may benefit from them for the duration of your lease only. If the capital improvement outlasts your lease, the next lessee, and not you, will obtain the benefits. The difficulty in agreeing upon capital improvements (who pays, who benefits, and so forth) tends to inhibit such improvements on leased boats.

Many fishermen anticipate the financial gains that can be made from buying and selling boats frequently. Such gains are not realized when a boat is leased. This is a disadvantage of leasing if financial gains are possible, but an advantage if buying and selling boats were to result in financial losses.

Leasing is undesirable to many fishermen because it connotes a

lack of financial success. The boat owner is one rung higher on the
ladder of success and prestige than the "hired skipper" — at least in
the minds of many fishermen. This feeling may prevail in spite of
evidence to the contrary.

If the lease is from a sea-food processor or dealer, there may be
another important disadvantage: the lessor (the processor or dealer)
may require your loyalty. You may be required to sell all of your
fish to and buy all your supplies from the lessor. You may receive
lower prices and pay higher prices than you would with other processors or dealers. It is an additional, though unpredictable, cost of
leasing.

Leasing Arrangements

No general recommendation can be made concerning leasing
versus owning a fishing boat. The decision depends upon your objectives, your financial situation, the characteristics of the fishery,
the availability of boats, and the type of lease that can be arranged.
Unfortunately, many lease arrangements are based upon tradition
and fail to remedy some of the disadvantages of leasing.

In any lease agreement, each party should share in gross returns
in exactly the same proportion as the party provides inputs to the business. Although this principle may appear abstract, it is basic to any
lease agreement. The interpretation of "gross returns" is obvious.
However, the interpretation of "inputs" deserves some attention.
The inputs to a fishing business can be classified as *labor*, *management*, and *capital*. The lessee usually provides most of the labor input, some of the management input, and very little of the capital
input (except for variable costs). The lessor may provide capital in
the form of a boat, gear, and fixed costs; some management; and
probably no labor.

The proportions of each input should be measured in terms of
value — for example, the lessee's labor at $3.50 per hour, the lessor's
boat at 9 percent of the market value per year, and management at
10 percent of gross returns. If the value of the lessee's inputs is 60
percent of the total and the value of the lessor's is 40 percent of the
total, a fair lease arrangement would allocate 60 percent of the gross
returns to the lessee and 40 percent to the lessor.

The lease should always be in written form. Careful consideration should be given to lease termination, lease renewal, liability, un-

foreseen circumstances, the determined value of the lessee's and lessor's inputs, and the shares of gross returns. Each party is more likely to understand the agreement if the terms are worked out together and are written into the lease. Although unforeseen difficulties, such as a scarcity of fish, bad weather, or a personal financial crisis, will strain any lease arrangement, difficulties can be minimized with a well-understood written agreement.

There are four basic types of lease arrangement:
- 1. Gross share lease.
- 2. Cash lease.
- 3. Profit share lease.
- 4. Fixed share lease.

Gross Share Lease

In this lease, the gross receipts are divided between the lessee and lessor. The lessee usually pays all variable costs and some fixed costs, contributes his own labor, and manages the operation of the boat. The lessor contributes the boat and the fixed gear, manages the capital, and pays some of the fixed costs.

The advantages of this type of lease are simplicity and the sharing of good as well as bad seasons (a sharing of risk). Gross receipts are easily calculated, and there is therefore little chance for disagreement when they are allocated to the lessee and the lessor.

The primary disadvantage of the gross share lease is the lack of incentive for greater fishing effort. Once all the fixed costs have been paid, the rational lessee will expend additional fishing effort only to the point where the addition to his share of the gross receipts is the same as the increase in his costs. The hypothetical example in Table 14-1 illustrates this problem. Suppose 100 days of fishing result in $10,000 in gross returns and $7,350 in variable costs, and 110 days of fishing result in $11,000 in gross returns and $7,700 in variable costs. The additional 10 days of fishing add $350 to costs and $1,000 to gross returns. The additional 10 days of fishing from 110 to 120 days are more difficult than the additional 10 days from 100 to 110 because of the need to fish in rougher weather, longer running times, and less time for preventive maintenance. Therefore, variable costs increase by $400 instead of $350, and gross returns increase by another $1,000 to $12,000. However, this is still a good move because gross returns again increase more than variable costs. Profit is still increasing for the fishing business.

TABLE 14-1. Hypothetical gross returns, costs, and profit for a lessee and lessor under a gross share fishing lease.

Total Fishing Days	Gross Returns	Increase in Gross Returns (Marginal Returns)	Variable Costs	Increase in Variable Costs (Marginal Costs)	Business Profit	Lessee's Share of Gross Returns	Increase in Lessee's Share of Gross Returns
100	$10,000		$ 7,350		$2,650	$5,000	
		$1,000		$ 350			$500
110	11,000		7,700		3,300	5,500	
		1,000		400			500
120	12,000		8,100		3,900	6,000	
		1,000		500			500
130	13,000		8,600		4,400	6,500	
		1,000		600			500
140	14,000		9,200		4,800	7,000	
		1,000		800			500
150	15,000		10,000		5,000	7,500	
		1,000		1,000			500
160	16,000		11,000		5,000	8,000	
		1,000		2,000			500
170	17,000		13,000		4,000	8,500	

In fact, increases in gross returns are greater than increases in variable costs up to 160 fishing days, the point at which profit is maximum. From 160 to 170 days, variable costs increase by $2,000 and gross returns increase by $1,000. This reduces profit from a maximum of $5,000 to $4,000.

Suppose you are leasing this boat and you pay all variable costs and receive 50 percent of gross returns. How many days would you fish? You would not fish more than 130 days. If you fished an additional 10 days beyond 130, your gross returns would increase by only 50 percent of $1,000, or $500, but your costs would increase by $600. Your own profit would decrease while the profit for the fishing business (and the boat owner) would still be increasing. This would obviously create a conflict between the lessee and the lessor, and would reduce the productivity of the boat (the capital invested). This is the major problem with the gross share lease.

Cash Lease

In the cash lease, the lessee and lessor agree upon some rental rate for the boat and gear, and the lessee pays the fixed rent regardless of his fishing success or lack thereof. The lessee again pays all variable costs, contributes his own labor and management, and may pay some fixed costs and provide some gear.

The advantages of this type of lease include simplicity, the likelihood that the boat will be better utilized, and fewer causes of conflict between lessee and lessor. The lease agreement must spell out the rental rate and the conditions of payment. Who is to pay what fixed costs and how capital improvements are decided upon and paid for should also be included.

We saw that, according to Table 14-1, the lessee was not likely to use the boat to its maximum profit potential if a percentage of gross returns were turned over to the lessor. This is not a problem with the cash lease arrangement because the cost of using someone else's boat and gear is a fixed cost, and the lessee will fish up to the point where the increase in *total* gross receipts is the same as the increase in variable costs — he will act the same as if he were an owner/operator.

Because the lessor or owner is guaranteed payment for the use of his boat and gear, he has little concern for how the lessee operates the business, provided there is no undue risk or damage to his property. Because of this, most of the possible causes of conflict in the lease arrangement are avoided. The lessor can relax and let the lessee worry about catching fish and obtaining a profit.

The primary disadvantage of the cash lease is that the lessee assumes all the risk of a bad season. The cash lease must be paid whether or not fish are caught. However, if the season is above average and the lease rate was based upon an average season, the lessee has the advantage. The other disadvantage of the cash lease is that it will be more difficult for the lessee to interest the lessor in making capital improvements to the boat. The only way the lessor can benefit from capital improvements is by increasing the lease rate. The lessee may benefit through increased landings and profit.

Profit Share Lease

In the profit share lease, the business is conducted much like a partnership: both lessee and lessor contribute something to the business, profits are calculated for the fishing business regardless of who pays what costs, and the profits are then divided between lessee and lessor according to their individual contributions. The primary advantage of this arrangement is that there is incentive for the lessee

TABLE 14-2. Hypothetical gross returns and lessee's and lessor's shares, under fixed share and gross share leases.

Season	Gross Returns	Fixed Share Lease [1]		Gross Share Lease [2]	
		Lessee's Share	Lessor's Share	Lessee's Share	Lessor's Sha
1972	$12,200	—	—	$6,100	$6,100
1973	5,100	—	—	2,550	2,550
1974	7,800	—	—	3,900	3,900
1975	10,500	$ 6,317	$4,183	5,250	5,250
1976	6,000	2,100	3,900	3,000	3,000
1977	18,000	13,950	4,050	9,000	9,000

[1] Fifty percent of previous three years' average annual gross returns to lessor; remainder to lessee.

[2] Fifty percent of current gross returns to both lessee and lessor.

and the lessor to manage the business for maximum profit. Criteria for business decisions are the same as those in an owner-operated fishing business.

The primary disadvantages of profit sharing are complexity and the necessity for both the lessee and lessor to be involved in many of the decisions and to agree upon the method of calculating profit. It is important that the lessee and lessor have access to each other and communicate on a businesslike basis. There will be many decisions that must be made on short notice and that cannot be anticipated in the lease agreement.

The profit share lease works well for the beginning fisherman and the retiring boat owner, for the shared management decisions provide an excellent opportunity for the novice to learn from the experienced boat owner. Also, the risk is shared equally.

The Fixed Share Lease

With the fixed share lease, the lessor's share is based upon average production or gross returns, not including the year of the lease. The share percentage is agreed upon in the lease, and the dollar amount allocated to the lessor is calculated by applying this percentage to the past two, three, or four years' gross returns. This becomes a fixed rent for the current season, and the lessee retains any gross returns in excess of this amount.

This arrangement combines the advantages of the cash lease (stability for the lessor) and the gross share lease (simplicity). However, the lessee assumes more risk than the lessor. This is illustrated in Table 14-2, which compares the lessee's and lessor's shares under fixed and gross share agreements. In this hypothetical fishery, gross returns vary from $5,100 to $18,000. If you wish to rent a boat (if you are the lessee) or if you have a boat to rent (if you are the lessor) for the 1977 season, you would calculate shares as follows:

1. Gross share lease: at the end of the 1977 season the gross returns are split equally, $9,000 to the lessor and $9,000 to the lessee.
2. Fixed share lease: at the beginning of the 1977 season you determine the lessor's share by calculating the average annual gross returns for 1974, 1975, and 1976, and multiplying this figure by 50 percent ($7,800 + $10,500 + $6,000, divided by 3 years, equals $8,100; $8,100 times 50 percent equals $4,050). The lessee's share is the 1977 gross returns ($18,000) less the lessor's share ($4,050) or $13,950.

With the gross share lease, the shares for the lessee and lessor vary equally from season to season. With the fixed share lease, the lessor's share varies less than the lessee's share. Also, the lessee benefits more from a good season and the lessor suffers less from a poor season.

Perhaps the greatest advantage to this type of lease is that both lessee and lessor benefit from good management and successive years of improvements and lease agreements. This advantage, plus the stability of income for the lessor, makes the fixed share lease attractive to the lessor who has unreliable alternative income sources and a long-term productive lessee.

Costs, Returns, and Boat Size

Your boat, whether owned or leased, is an input to your fishing business. In Chapter 6, we saw how inputs were related to costs, returns, and profit. The same relationship applies to your boat; the larger the boat, the greater the costs, returns, and profit *up to a given point*.

How do we measure boat size? Length may be important in one fishery, capacity in others. Speed may be important in some fish-

TABLE 14-3. A hypothetical season of shrimp production: total costs and total costs per ton for eight boat sizes, showing economies and diseconomies of size.

Boat Size	Season Production in Tons	Total Costs	Total Costs Per Ton
Rowboat and hand net	0.5	$ 1,200	$2,400
Power skiff and hand net	1	1,800	1,800
30-ft displacement hull, 1 power net	5	7,000	1,400
50-ft displacement hull, 2 nets	20	24,000	1,200
70-ft displacement hull, 2 nets	40	40,000	1,000
100-ft displacement hull, 2 nets	50	70,000	1,400
140-ft displacement hull, 3 nets	58	116,000	2,000
180-ft displacement hull, 4 nets	65	182,000	2,800

eries, and seaworthiness is important in others. The dollar value of boats can also be used as a measure of size, but this is not without problems. How is dollar value calculated? Original cost is one approach, but construction costs have recently increased dramatically. Market value is another approach, but this measure reflects the economic condition of the fishery more than boat size.

There is no one best measure of boat size. A combination of the above measures, along with special recognition for characteristics important in the particular fishery, is a good compromise.

Regardless of the measure of size used, it is still possible to determine the optimum boat size for any fishery. This is the size at which the total costs per unit of fish produced are lowest.

Decreasing Cost

Beginning with a small boat, one fishery, and any one season, the cost per unit of fish produced is lower for larger boats. This is referred to as economies of size. Economies of size are possible because larger boats are more efficient. They can haul more and larger gear with less fuel. They can reach the fishing grounds sooner, stay longer, and return sooner than smaller boats. They can carry more fish each trip and handle the fish more efficiently.

This principle is clearly illustrated in Table 14-3. Suppose you have chosen to go shrimp fishing next year and are considering the purchase or lease of a boat. To illustrate our point, you could use a rowboat with a hand net. Your total costs would be very low, but your catch would likely be even lower. This would give you a very high cost per ton. From Table 14-3, we can see that rowboat costs are $2,400 per ton. If you move up to a powered skiff, your shrimp catch increases more than your total costs and shrimp costs go down to $1,800 per ton. If you move up to a 30-foot displacement-hull boat with a powered net, shrimp catch again increases more than total costs and costs per ton drop to $1,400. The cost per ton continues to drop as you move to larger and larger boats, up to the 70-foot shrimper with two shrimp nets.

Economies of size are possible because the powered skiff can drag the net faster than the rowboat, the powered net can be hauled and set faster than the hand net, the displacement-hull boat can hold more shrimp per trip, and the larger boat can travel farther and fish longer in rougher weather than the small boat. The technical advantages of the larger boat more than offset its greater total cost.

Increasing Costs

For one fishery and any one season, the cost per unit of fish will increase as boat size increases. This is called diseconomies of size. This is also illustrated in Table 14-3. The 70-foot boat with two nets can land shrimp for $1,000 per ton. Moving up to a 100-foot boat increases the maximum possible catch from 40 to 50 tons, but this increase isn't sufficient to offset the increased costs of owning (or leasing) and operating such a large shrimp boat. Although its hold may be larger, it may not be possible to fill it very often, given the amount of shrimp available. Although the larger vessel can stay at sea longer than the smaller one, it may be forced to return as frequently as the 70-foot boat to deliver shrimp of equal quality. It will take a 100-foot shrimper as long to search out new shrimp grounds as a 70-foot boat, but the cost of operating the 100-foot boat while searching will be greater. Therefore, costs per ton for the 100-foot boat are $1,400, which is greater than the $1,000 cost per ton for the 70-foot boat.

If this is the case (in our hypothetical example) for a 100-foot shrimper, it is also true for a 140-foot shrimper and a 180-foot shrimper. The costs for these larger boats are proportionately higher than the shrimp catch. Total cost per ton increases rapidly as we move up from a 70-foot boat to a 180-foot boat.

Making the Boat-Size Decision

There are four primary factors to consider in deciding upon boat size. Not necessarily in order of importance, they are:

— 1. Characteristics of the fishery.
— 2. Your mobility.
— 3. Your financial situation.
— 4. Economies and diseconomies of size.

Each of these factors has been discussed in previous chapters. However, it will be useful to review them in relation to the boat size decision.

Characteristics of the Fishery

This includes such things as: the availability and seasonal variation in harvestable fish stocks, type and predictability of fishing regulations, type and predictability of weather and oceanic conditions, distance to market, variability and predictability of market price, and supply of labor, gear, fuel, and so forth. These characteristics were discussed in Chapter 13. The greater the variability in any of the above characteristics, the greater the risk.

If the primary source of risk is variability in weather, a larger boat will offset the economic impact of such variability. You will be able to fish longer if and when you experience rough weather.

If the primary sources of risk are variability in fish stocks, regulations, and prices, a smaller boat provides a hedge against diseconomics of size. However, the ultimate decision will depend upon your decision criteria. For example, if you expect the best, you will decide in favor of a larger boat. If you expect the worst, you will opt for a smaller boat. If you expect and can calculate the average fish stock, regulations, weather, and so forth, you will select the lowest-cost-per-ton boat size.

Your Mobility

If you anticipate changing fisheries frequently or moving in and out of the fishing business frequently, you may need to consider the marketability of the boat in determining its size. A smaller boat is frequently easier to sell than a larger boat. The smaller the equity requirement in purchasing a boat, the greater the number of potential customers. There are more people with a little capital than there are people with a lot of capital.

Potential economic losses in attempting to sell a larger boat (with the lowest cost per ton) may offset the potential gain from economies of size. The more frequently you buy and sell, the more important becomes the marketability factor relative to economies of size.

Your Financial Situation

The ultimate limit to boat size lies in your ability to finance the purchase or lease of a boat. The old adage, "It takes money to make money," applies here. The least-cost-per-ton boat may not be attainable if your equity and borrowing capacity are too small. The dangers of overinvesting in a boat you own were pointed out earlier in this chapter. Once you have determined the optimum boat size for you, you must analyze your own financial situation and determine whether that boat size is financially attainable. What size boat is possible given your financial situation, and will it produce a profit so that equity can be increased and the optimum size eventually purchased? Should you look at another fishery, one in which the optimum boat size is more compatible with your financial situation? Should you consider a lease arrangement, a partnership, or incorporation? These are all reasonable alternatives and should be kept in mind when deciding on boat size.

Your Management

We have discussed most of the factors that determine economies and diseconomies of size. The most important factor of all is your own management. Some fishermen have the management and fishing ability to sustain minimum costs per ton from a $1,000,000 tuna seiner, but others are unable to make any profit from a 40-foot lobster boat. Your management capacity must match the capacity of your boat; otherwise, you may suffer diseconomies of size with the smallest boat in the fleet!

Use the decision-making steps outlined in Chapter 2 in deciding upon the optimum boat size:

1. Gather ideas and information: select your fishery and accumulate as much costs and returns data as possible on different boat sizes in that fishery.
2. Analyze the information: calculate costs per ton for various levels of production for each boat size. Find the lowest-cost-

per-ton production level for each boat size and compare them. Prepare projected financial statements and projected profit and loss statements based on the purchase (or lease) of the lowest-cost-per-ton boat.

3. Make the decision.
4. Implement the decision.
5. Take responsibility for the decision: keep good records and compare them with your projected financial statements and projected profit and loss statements.

15   MANAGING THE BOAT

Parts I and II of this book are devoted to management concepts and tools. Chapters 13 and 14 demonstrated the application of these concepts and tools to three infrequent but very important decisions: selecting a fishery, boat size, and boat ownership. In this chapter, we will illustrate the application of management concepts and tools to the more frequent operational decisions. Although each operational decision may be less important than selecting a fishery or a boat, their cumulative impact upon profits is probably greater.

Use of Labor and Capital in the Fishing Business

All inputs to the fishing business can be classified as labor, capital, or management. Your decision making represents management. Labor is represented by the number of employees and the hours they devote to your business, plus your own investment of time. Capital is represented by your boat, fishing gear, deck gear, electronics, refrigeration, engines, and total costs.

There are many combinations of labor and capital that will produce the same gross returns. Labor will substitute for capital or capital will substitute for labor in painting a boat, hauling gear, unloading fish, repairing gear, replacing an engine, and so forth. In deciding what combination of labor and capital to use, the following two principles can be applied: (1) substitute labor for capital or capital for labor as long as costs are reduced and results are not changed, or (2) substitute labor for capital or capital for labor as long as results are improved and costs do not change.

TABLE 15-1. Hypothetical labor costs, capital costs, and total costs for three methods of painting a boat, illustrating labor and capital substitutions.

Cleaning and Painting Method	Labor Costs	Capital Costs	Total Costs
Three men cleaning with chisels and brush and hand-painting	$420	$195	$615
Two men cleaning with chisels and brush and spray-painting	320	285	605
One man cleaning with explosives and chemicals and spray-painting	260	365	625

Labor/Capital Substitution with Results Constant

Suppose you have a steel boat that is due for a complete painting. You can hire three men to clean the boat with chisels and wire brushes and paint it by hand. You can hire two men to clean with chisels and wire brushes and paint with a pressure sprayer. Or you can hire one man to clean with explosive net charges and chemicals, and paint with a pressure sprayer. The labor costs, capital costs, and total costs of each labor/capital combination are illustrated in Table 15-1.

From Table 15-1, we can see that the first method has a total cost of $615. The labor cost is less with the second method because we substitute a power sprayer for a man with a paint brush, but the rental of the sprayer increases capital costs. Nevertheless, the total costs of the second method are lower than those of the first method. Finally, we can hire a specialist to clean our steel hull with a specially designed explosive net and other chemicals. Because this takes him only one day, labor costs are again reduced. However, the explosive net and chemicals raise capital costs again, and total costs are greater than with the first two methods.

The best combination of labor and capital is the second method — two men with chisels and wire brushes and a third using a power paint sprayer — provided the resulting paint job is equal to that resulting from the other two methods.

Labor/Capital Substitution with Costs Constant

For another illustration, suppose you have $4,000 available annually and wish to improve the fish-keeping capacity of your dragger.

You can accomplish this in one of several ways:
1. Employ another man, who will use a shovel and keep the fish and ice stored properly.
2. Purchase an "on-deck" ice-making machine and assign a part-time man to ice and store fish.
3. Purchase a complete on-board refrigeration system that will automatically store and ice the fish.

The first method allocates nearly all the $4,000 to labor. The second method allocates part of the $4,000 to labor and part of it to capital. The third method allocates the entire $4,000 to capital. The method yielding the best result should be the method used, because costs for all three methods are equal.

Table 15-2 illustrates the different results from these three methods. With the first — having a man with a shovel in the hold and taking on ice at the fish dock — an average of 18 tons of fish are landed per trip. Capacity is limited by the necessity of having a man in the hold (he needs room to shovel) and space for the ice, whether or not all the ice is needed. Also, trips are shortened by loss of ice, reducing the number of fish landed per trip. Because the fish are moved about the hold, sometimes into bilge water, quality and therefore price is lower. There are more trips per season due to the shorter trip length. The season's gross returns are $174,240.

The second method, an "on-deck" ice machine and a part-time man in the hold, makes it possible to carry more fish (less shoveling room needed and no "extra" ice needed), stay at sea longer, and land a product of higher quality. Longer trips result in fewer trips, and season's gross returns are $170,430.

TABLE 15-2. Hypothetical results from three methods of increasing fish-keeping capacity aboard a dragger, illustrating labor and capital substitution.

| Method | Labor Cost | Annual Capital Cost | Total Cost | Results | | | | |
				Tons of Fish Per Trip	Price Per Ton	Gross Returns Per Trip	Trips Per Season	Season's Gross Returns
Man with shovel	$3,988	$ 12	$4,000	18	$220	$3,960	44	$174,240
Ice Machine	1,600	2,400	4,000	19	230	4,370	39	170,430
Refrig- eration	0	4,000	4,000	27	250	6,750	33	222,750

The third method allows maximum capacity in the hold, top-quality fish, and therefore top price. The $222,750 in gross returns is the highest. Therefore, all of the available $4,000 should be allocated to a complete refrigeration system (capital) and none to labor.

The foregoing illustrations are simplified to make the principle clearer. However, they contain some assumptions that deserve comment. In the first illustration, it is assumed that the labor must be hired. For many fishermen it is understood that boat repair and maintenance is part of labor's responsibility, and no extra pay over and above the normal crew-share would be required for chipping and painting. If this were the case, the first method would result in the lowest cost. In the second illustration, it is assumed that you have the financial capability to buy an ice machine or a complete refrigeration system. The annual capital cost of $4,000 is the annual interest, depreciation, and upkeep of the capital asset and may have to be sustained for ten or more years. Finally, fewer trips with more tons of fish per trip will result in lower costs per ton for the business as a whole. This may also enter into the decision, but it would not change the result in our example.

The Management of Labor

The relationships between crew and skipper and crew and boat are special in the fishing business. The crew's livelihood as well as its life and welfare depend upon the skipper and the boat. Therefore, safety, health, and social considerations are as important as economic considerations in labor management. However, safety, health, and social considerations are beyond the scope of this book.[1]

Rate and Method of Pay

Most employees in the fishing industry are paid a share of the gross returns. Employees are referred to as "boat pullers," "crewmen," "sternmen," "deck hands," and so forth, and their pay is referred to as the "lay" or "crew-share." There are almost as many rates and methods of pay as there are fisheries. In some fisheries,

---

[1] An excellent discussion of these considerations can be found in the *Marine Fisheries Review*, Volume 36, Number 6, National Oceanic and Atmospheric Administration, National Marine Fisheries Service.

certain costs are first deducted from gross returns and a percentage of the remainder is divided among the employees. In other fisheries, the employees divide a share of gross returns before any deductions. Some fisheries pay employees a certain amount per unit of fish landed, less some variable costs.

In all cases, employees share some of the economic as well as the physical risk of fishing. The boat owner's and/or operator's risk is thereby reduced. This share system also provides an incentive to employees for greater productivity. In fact, share systems that require employees to pay part of the variable costs are designed to temper the overzealous employee who would increase production at any cost.

Your rate and method of pay will depend upon two major factors: (1) the supply of labor, and (2) the employee's contribution to your gross returns and profit. If there is a shortage of labor, you will have to pay at least the prevailing rate. Your reputation and the condition of your boat will be very important subsidiary factors in attracting employees. If you have a reputation as a very productive but fair skipper, the percentage of gross returns becomes less of a factor in hiring. If you maintain a safe, seaworthy, and comfortable boat, hiring will be easier.

If there is an abundance of labor, the second pay factor becomes more important than the first. How can you measure an employee's contribution to gross returns? You can only guess at this before you hire. It is important to evaluate the potential employee as thoroughly as possible before hiring. Many fishermen evaluate the purchase of a fender or a stay more carefully than a potential employee, yet an employee will have a much greater impact upon profit (and safety). Find out where the prospective employee has worked before. Call his former employer and determine why he is no longer employed by him. What are his strengths, weaknesses, and special skills? Interview the potential employee and determine why he wants to work on a boat and why he wants to work for you. What are his long-range plans?

It is easier to measure an employee's contribution to gross returns after he has worked for you. Suppose you own a deepwater lobster boat and have added a fifth man to your crew. You are paying the new man one fifth of the total crew-share, or 7.3 percent of the gross returns. If the addition of this fifth man increases gross returns by the same dollar amount that he receives, you are paying exactly what he contributes. This doesn't necessarily mean that you

TABLE 15-3. Hypothetical returns and costs with one, two, three, four, and five employees o
deepwater lobster boat.

Number of Employees	Gross Returns	Marginal Returns	Percentage of Total Share	Total Crew-share	Marginal Costs	Crew-sh per Emp
1	$15,000		20	$3,000		$3,00
		$5,000			$2,000	
2	20,000		25	5,000		2,50
		3,000			1,900	
3	23,000		30	6,900		2,30
		2,000			1,600	
4	25,000		34	8,500		2,12
		1,000			1,000	
5	26,000		36.54	9,500		1,90

should have hired the fifth man. The optimum number of employees
is found by comparing the increase in gross returns (marginal returns)
with the increase in labor costs (marginal costs).

This relationship is demonstrated in Table 15-3. With one employee, gross returns are $15,000 and his share is $3,000. With two
employees, gross returns jump to $20,000, total crew-share is $5,000,
and each employee receives $2,500. Gross returns continue to in-
crease at a decreasing rate as employees are added. When the fifth
employee is added, gross returns increase by $1,000 (marginal re-
turns) and labor costs increase by $1,000 (marginal costs). Five em-
ployees are the optimum number, but the fifth employee is receiving
wages greater than the increase in gross returns. His share of the
crew-share is $1,900, but the marginal return is $1,000. This is because shares are divided equally among all employees, whether they
are hired first or last. When the third employee was added he received $2,300, but gross returns increased by $3,000. The third employee was "underpaid," whereas the fourth and fifth were "overpaid."

Employee Benefits

Increased capital requirements in the fishing business have increased the importance of steady, well-trained employees. The casual
crewman can be very costly, in spite of a low pay rate, if gear, hydraulic systems, or electronics are damaged. Both you and your employees benefit from a reliable, well-trained crew. This is accomplished through careful hiring practices, good employee-employer relationships, and an on-the-job training program.

Other fringe benefits for employees include your contributions to social security, workman's compensation, health care, and other insurance plans. The retention and loyalty of competent employees usually offset the small cost of such fringe benefits.

Employee-Employer Relationships

In addition to reasonable pay rates and fringe benefits, an understanding of employee responsibilities and the application of good human-relations principles are necessary to attract and keep employees. You should make clear just what you expect of an employee in return for his crew-share. You may feel that any and all of an employee's time is at your command. The employee may feel that his responsibility is limited to the normal fishing activities at sea and that his time is his own when he is in port. This type of misunderstanding contributes to many employee-employer problems.

Any fisherman who has one or more employees will benefit from a study of human-relations principles. Your personal treatment of an employee may compensate for low pay, lack of fringe benefits, and hard work, or it may cause an excellent employee to quit in spite of all monetary benefits. There are many human-relations principles, but perhaps the most important is to recognize an employee for his contribution to your fishing business. Give him honest praise when he earns it.

The Management of Fishing Gear

Some of the most frequent fishing business decisions relate to gear. What type of gear should I use? How much gear should I own? When should I replace or repair my gear? We will illustrate each of these decisions by using management tools developed earlier.

What Type of Gear

There are several types of gear that can be used in any one fishery. Some of the alternatives are restricted by regulations, but within a fishery and within the regulations you can decide between wood, wire, or plastic traps in lobster fishing; single net drum or twin briole nets in the shrimp fishery; long lines or pots in the black cod fishery; and so forth. In addition to the various types of fishing gear avail-

TABLE 15-4. Hypothetical partial budget for converting from hand tongs to hydraulic-power tongs on an oyster boat.

Increased Costs

Repair and maintenance	$920	
Fuel	308	
Depreciation	520	
Opportunity cost of investment	286	
Total increased costs		$2,034

Decreased Costs

Repair and maintenance	$120	
Food	165	
Labor	840	
Total decreased costs		$1,125

Increased Receipts

1200 bushels at $4.10 per bushel	$4,920	
Total increased receipts		$4,920

Decreased Receipts

None	$0	
Total decreased receipts		$ 0
Decreased costs plus increased receipts		6,045
Less increased costs plus decreased receipts		1,125
Net benefit		+ $4,011

able, there are various types of auxiliary deck gear that can be used.

Suppose you own and operate an oyster boat and employ two men who use hand tongs for harvesting. You are considering a conversion to hydraulic-powered oyster tongs. We will follow the decision-making steps outlined earlier in making this decision.

First, you must determine how costs and returns will be affected if you convert to hydraulic-power tongs. Interviews with oyster fishermen who use these tongs will provide most of the information you need:

1. The hydraulic system (including auxiliary engine and tongs) will cost $2,600 and has an expected useful life of five years.
2. The opportunity cost of capital is 11 percent.
3. Production would likely increase by 1,200 bushels per season.
4. Hydraulic tongs require one less crewman than hand tongs.

From this and other information we can prepare a hypothetical partial budget (Table 15-4). The new gear will require additional repair and maintenance, estimated to be $920 annually. The auxiliary engine will require $308 worth of fuel. The hydraulic system, and tongs will be replaced in five years, giving an annual depreciation of $520. Finally, the opportunity cost of 11 percent times the $2,600 initial investment will be $286 the first year. Total increased costs are $2,034.

Decreased costs in Table 15-4 are $120 for repair and maintenance of hand tongs no longer used, for the $840 crewman who would be discharged, and $165 for the food that would not be eaten by this person. Increased receipts are from 1,200 additional bushels of oysters harvested at $4.10 per bushel.

If the boat is converted to hydraulic-power tongs, the partial budget tells us that you have gained $4,011 over and above previous earnings. Whether you convert to hydraulic tongs will also depend upon several nonmonetary factors: Will hydraulic tong regulations become more stringent? Will the available supply of oysters decline? Will other uses of the boat become more attractive? What will happen to the discharged crewman?

Are any of these nonmonetary factors significant enough to outweigh the potential $4,011 net benefit? If not, then the decision is yours to make and implement. Once implemented, you must also assume responsibility for the results.

How Much Gear?

It is not uncommon for a gill-net fisherman to own six, seven, or even eight nets and use only one of these nets in several successive seasons. Although this may look like gross overinvestment, there are some economic reasons for owning "extra" gear. Of course, there are compelling reasons not to own "extra" gear as well. In fisheries where a fixed amount of gear can be fished at any one time, there are three important reasons to own extra gear:

1. To minimize nonfishing time when the gear being fished is damaged or lost.
2. To allow for flexibility when changes in fishing conditions make the gear being fished nonoptimal.
3. To provide a hedge against rising gear prices or an unreliable supply.

In fisheries where the amount of gear is not fixed (pot, long-line,

TABLE 15-5. Hypothetical budgeted costs for retaining old refrigeration system, installing new refrigeration system in 1977, and installing new refrigeration system in 1978.

	Old System			
Year	Maintenance & Repair	Depreciation	Opportunity Cost	Total
1976	$ 690	$600	$270	$1,560
1977	784	600	216	1,600
1978	918	600	162	1,680
1979	1,072	600	108	1,780
1980	1,276	600	54	1,930

	New System in 1977			
Year	Maintenance & Repair	Depreciation	Opportunity Cost	Total
1976	—	—	—	—
1977	$ 100	$800	$720	$1,620
1978	120	800	648	1,568
1979	170	800	576	1,546
1980	260	800	504	1,564

	New System in 1978			
Year	Maintenance & Repair	Depreciation	Opportunity Cost	Total
1976	—	—	—	—
1977	—	—	—	—
1978	$ 100	$800	$720	$1,620
1979	120	800	648	1,568
1980	170	800	576	1,546

and trap fisheries), there is a fourth reason: extra gear allows for immediate expansion of fishing capacity when fishing conditions warrant.

The first reason is particularly important if there is no on-board capability for repairing or retrieving gear. In most fisheries, net mending is an important but disappearing skill. An extra net on board will substitute for a slow, non-net-mending crew. The extra net will increase fishing time per season, particularly if you fish multi-day trips several hundred miles from port. In fisheries that are closely regulated, an extra net may make the difference between 15 and 40 tons of production. The salmon gill-net fishery in the

Pacific Northwest is a prime example. Some seasons are only five days long. A torn gill net on the first day (partly submerged logs and sturgeon make short shrift of gill net) may beach the gill netter for the season — unless he has an extra net.

The second reason is important in fisheries where fishing conditions change and are unpredictable. The Rhode Island lobster fisherman with only 500 traps will not have as much flexibility as one with 800 traps if the lobster production drops and red crab production jumps overnight. The "extra" traps will allow you to set quickly in the red crab area before hauling and moving your 400 or 500 traps out of the lobster area.

The third reason for owning extra gear has some flaws. Holding gear because you expect it to cost more next year is speculating with your own capital. Unless you are an expert on the "gear market," there is probably less risk if you invest that capital in some other way.

The added cost of owning extra gear is not difficult to calculate. It is the opportunity cost of the capital invested, storage costs, depreciation, and any repairs that may be required. The added benefits are more difficult to calculate — at least in dollars and cents. If an extra net on board allows you to complete one trip during the year with a full load instead of a half load, the dollar benefit will be the value of that half load. If an extra 300 traps allows you to land 8 more tons of red crabs in 2 days of fishing, that is the dollar benefit of the 300 traps. Other benefits can be calculated, but it will require good log information, an accurate diary, and complete records. Nevertheless, it is important to calculate the dollar benefits and to make the "extra" gear decision on a dollars-and-cents basis.

Gear Replacement and Repair

Next to labor, gear repair and replacement constitute the most costly part of a fishing business. As fishing gear increases in age, the repair and maintenance costs increase even faster. At some point in time, the annual cost of new gear will be less than that of old gear.

We can illustrate this by comparing the annual repair and maintenance of an old refrigeration system with that of a new refrigeration system. Table 15-5 illustrates the old versus new refrigeration alternatives and the relevant costs for five years. In 1976, the maintenance and repair costs, the annual depreciation, and a 9 percent opportunity cost of the remaining refrigeration system value (opportunity cost) totals $1,560. Due to greatly increased maintenance and repairs, this jumps to a projected $1,600 in 1977. Should we trade in the old system and buy a new one at the beginning of 1977?

The annual cost of the new system in 1977 is greater than that of the old one — $1,620 versus $1,600. If we wait until the end of 1977, we see that the 1978 cost of the old system exceeds the 1978 cost of the new system — $1,680 versus $1,620. We should trade in the old refrigeration system and install the new one no later than the beginning of 1978. However, by waiting until the beginning of 1978 we have foregone the even smaller second-year annual cost of the new system installed in 1977. We can see this by adding the 1977 and 1978 total costs for all three alternatives and comparing:

Total Annual Costs

	1977	1978	Two-Year Total
Old system	$1,600	+ $1,680 =	$3,280
New in 1977	$1,620	+ $1,568 =	$3,188
New in 1978	$1,600	+ $1,620 =	$3,220

The "new-in-1977" alternative continues to improve in 1979 because total costs continue to decline.

This is a fairly simple treatment of a complex problem. The procedure can be applied in any old-versus-new, repair-versus-replace, and rent-versus-buy decision in which costs and/or returns will continue for several years. It is not sufficient to base your decision on a single year's projection. Neither is it sufficient to base your decision upon intuition.

Determinants of Profit

Gross returns less costs yields profit. Increases in gross returns over time (several years) generally contribute more to profit than do decreases in total costs. Gross returns equals production times price. Because you are likely to have little influence upon price, you should concentrate upon production.

Production depends upon two factors:

— 1. The harvestable fish available.
— 2. The amount, quality, and combination of inputs to your business.

Because you have little control over the harvestable fish available, you should concentrate on the inputs to your fishing business. The most important inputs to your fishing business are your management, your fishing skill, and total fishing effort.

Fishing effort comprises the amount and type of gear you have in the water, the number of hours or days the gear is in the water, the gear-handling capabilities of your boat and crew, and the fish-carrying capacity of your boat. Therefore, you should allocate your scarce management time to improving fishing gear and boat productivity. Improvements in gear and boat should be made as long as the increase in gross returns (marginal returns) is as large as the increase in costs (marginal costs).

This is not to say that total costs are unimportant or that other inputs do not influence profit. The primary determinants of profit may vary in importance from fishing business to fishing business and from season to season. Identify what is important in your fishing business, and follow these rules for more effective management:

1. Have your objectives well in mind.
2. Maintain a good information system.
3. Use a scientific decision procedure.
4. Consider monetary and nonmonetary factors separately.
5. View your fishing business as a business and yourself as a manager as well as a fisherman.

BIBLIOGRAPHY

BOOKS

Alpin, R. D., and George L. Casler. *Capital Investment Analysis*. Columbus, Ohio: Grid Inc., 1973.

Bell, F. W., and J. W. Hazelton, eds. *Recent Developments and Research in Fisheries Economics*. Dobbs Ferry, N. Y.: Oceana Publications, 1967.

Brock, Horace R., Charles E. Palmer, and Fred C. Archer. *College Accounting —Intermediate*. American Institute of Banking Edition. New York: McGraw-Hill Book Co., Inc., 1967.

Castle, E. N., M. H. Becker, and F. J. Smith. *Farm Business Management*. New York: The Macmillan Co., 1972.

Donaldson, Elvin F., and John K. Pfahl. *Personal Finance*. 5th edition. New York: The Ronald Press Company, 1971.

Johnson, Robert W. *Financial Management*. Boston: Allyn & Bacon, Inc., 1966.

Klatt, L. A. *Small Business Management, Essentials of Entrepreneurship*. Belmont, Calif.: Wadsworth Publishing Company, 1973.

Van Voorhis, Robert H., Charles E. Palmer, and Fred C. Archer. *College Accounting Theory and Practice* (Parts 1 and 2). New York: McGraw-Hill Book Co., Inc., Gregg Publishing Division, 1963.

BULLETINS, REPORTS, AND MANUALS

Ahsan, A. E., J. L. Ball, and J. R. Davidson. *Costs and Earnings of Tuna Vessels in Hawaii*. The University of Hawaii Sea Grant Program, UNIHI-SEAGRANT-AR-72-01. July, 1972.

Anderson, C. L., and R. H. McNutt. *Costs and Returns in Commercial Fishing: Mullet Fishing – Florida, A Case Study*. Florida Cooperative Extension Service Marine Advisory Program, SUSF-SG-73-002, 1973.

Bell, Frederick W. *The Economics of the New England Fishing Industry: The Role of Technological Change and Government Aid*. Research report to the Federal Reserve Bank of Boston, No. 31. February, 1966.

Campbell, Blake A. *Returns from Fishing Vessels in British Columbia*. Office
  of Director, Pacific Region-Fisheries Service, Department of Fisheries and
  Forestry, 1155 Robson St., Vancouver 5, B.C. Issued periodically.
Carley, D. H. *Economic Analysis of the Commercial Fishery Industry of
  Georgia*. University of Georgia Agricultural Experiment Station, Research
  Bulletin 37. June, 1968.
Crutchfield, James A. *Pacific Northwest Economic Base Study for Power
  Markets: Fisheries*, Vol. II, Part 8. U.S. Department of the Interior,
  Bureau of Commercial Fisheries. 1967.
Gates, J. M., G. C. Matthiessen, and C. A. Griscom. *Aquaculture in New
  England*. University of Rhode Island, Marine Technical Report No. 18.
  Kingston, R. I.:1974.
Gates, John M., and Virgil J. Norton. *The Benefits of Fisheries Regulation:
  A Case Study of the New England Yellowtail Flounder Fishery*. University
  of Rhode Island, Marine Technical Report No. 21. Kingston, R. I.: 1974.
Green, Roger E., and Gordon C. Broadhead. "Costs and Earnings of Tropical
  Tuna Vessels Based in California." FIR Reprint 31 from *Fishery Industrial
  Research*, Vol. 3, No. 1, pp. 29-45. Bureau of Commercial Fisheries.
Greene, Mark R. *Insurance and Risk Management for Small Business*. U.S.
  Small Business Administration, SBA Series No. 30. Washington, D.C.:
  1970.
Holmsen, Andreas. *Remuneration, Ownership and Investment Decisions in the
  Fishing Industry*. University of Rhode Island, Marine Technical Report
  No. 1. Kingston, R. I.: 1972.
———. *Financing Fishing Vessels*. University of Rhode Island, Marine Memo-
  randum No. 34. Kingston, R. I.: 1974.
Hourston, W. R., and Blake A. Campbell. *You are Operating and Managing a
  Fishing Business*. British Columbia Department of Fisheries and Forestry,
  1155 Robson St., Vancouver 5, B.C., 1971.
Lacewell, Ronald D., Wade L. Griffin, James E. Smith, and Wayne A. Hayenga.
  Estimated Costs and Returns for Gulf of Mexico Shrimp Vessels: 1971.
  Texas A & M University, Texas Agricultural Experiment Station, Depart-
  ment Technical Report No. 74-1. January, 1974.
Merrill Lynch, Pierce, Fenner and Smith Inc. *How to Read a Financial Report*.
  December, 1968.
Norton, Virgil J., and Morton M. Miller. *An Economic Study of the Boston
  Large Trawler Labor Force*. U.S. Department of the Interior, Fish and
  Wildlife Service, Bureau of Commercial Fisheries Circular 248. Washington,
  D.C.: May, 1966.
Perrin, William F., and Bruno G. Noetzel. "Economic Study of the San Pedro
  Wetfish Boats," *Fishery Industrial Research*, Vol. 6, No. 3, pp. 105-138.
  U.S. Fish and Wildlife Service, Bureau of Commercial Fisheries. June, 1970.

Proskie, J. *Costs and Earnings of Selected Fishing Enterprises, Nova Scotia*. Economics Branch, Fisheries Service, Department of Fisheries and Forestry. Ottawa. Issued periodically.

Redfield, Michael. *Costs and Profitability in the Commercial Fishing Industry: The Insurance Dilemma*. University of Washington, Division of Marine Resources, WSG-MP 71-4. 1971.

Sielken, R. L., Jr., R. G. Thompson, and R. R. Wilson. *Extended Results on Optimal Investment Strategies in Shrimp Fishing*. Texas A & M University Sea Grant Program, TAMU-SG-72-211. December, 1972.

Smith, Frank J., Jr., and Ken Cooper. *The Financial Management of Agribusiness Firms*. University of Minnesota, Special Report 26 (September, 1967).

Smith, Frederick J. *Incorporating a Fishing Business*. Oregon State University, SG No. 11. Corvallis, Oregon: 1973.

————. *Fishing Business Management and Economic Information*. Oregon State University, SG No. 6. Corvallis, Oregon: 1973.

————. *Organizing and Operating a Fishery Cooperative*. Oregon State University, SG No. 19. Corvallis, Oregon: 1972.

————. *Understanding and Using Marine Economic Data Sheets*. Oregon State University, SG No. 24. Corvallis, Oregon: 1973.

————. *How to Calculate Profit in a Fishing Business*. Oregon State University, SG No. 29. Corvallis, Oregon: 1973.

Thompson, Russell G., Richard W. Callen, and Lawrence C. Wolken. *Optimal Investment and Financial Decisions for a Model Shrimp Fishing Firm*. Texas A & M University Sea Grant Program, TAMU-SG-70-205. April, 1970.

U.S. Department of Commerce. *Basic Economic Indicators*. Current Fisheries Statistics Number 5934 and Numbers 6127 through 6133. Economic Research Laboratory, National Oceanic and Atmospheric Administration, National Marine Fisheries Service. Washington, D.C.

————. *Marine Fisheries Review*, Vol. 36, No. 6. National Oceanic and Atmospheric Administration, National Marine Fisheries Service. Washington, D.C.: June, 1974.

U.S. Internal Revenue Service. *Tax Guide for Small Business*. Technical Publications and Services Division, Publication No. 334. Internal Revenue Service, Washington, D.C., 20224.

Wilson, Robert R., Russell G. Thompson, and Richard W. Callen. *Optimal Investment and Financial Strategies in Shrimp Fishing*. Texas A & M University Sea Grant Program, TAMU-SG-71-701. December, 1970.

Wilson, W. Alan. *An Analysis of Results from the Fishery Economic Survey*, 1970, British Columbia Department of the Environment, Fisheries Service, Pacific Region, Detail Reports Nos. 1, 2, and 3. Vancouver, B.C.: 1971.

Zwick, Jack. *A Handbook of Small Business Finance.* U.S. Small Business Administration, SBA Series No. 15. Washington, D.C.: 1965.

OTHER SOURCES OF INFORMATION

Alaska Marine Advisory Program, 102 Eielson Bldg., University of Alaska, Fairbanks, Alaska 99701

California Marine Extension Service, Department of Animal Physiology, University of California, Davis, Calif. 95616

Delaware Marine Advisory Service, Lewes Field Station, University of Delaware, Lewes, Del. 19958

Florida Marine Advisory Program, Room 1038, McCarty Hall, University of Florida, Gainesville, Fla. 32601

Georgia Marine Resources Extension Center, University of Georgia, P. O. Box 13687, Savannah, Ga. 31406

Hawaii Marine Advisory Service, University of Hawaii, 2540 Maile Way, Spaulding 235, Honolulu, Hawaii 96822

Humboldt State University Marine Advisory Extension Service, Arcata, Calif. 95521

Louisiana Sea Grant Program, Coastal Studies Building, Louisiana State University, Baton Rouge, La. 70803

Maine Cooperative Extension Service, University of Maine Marine Laboratory, Walpole, Maine 04573

Maine Department of Sea and Shore Fisheries, State House Annex, Augusta, Maine 04330

Maryland Cooperative Extension Service, 1215 Symons Hall, University of Maryland, College Park, Md. 20742

Michigan Advisory Service Program, 4117 Institute of Science and Technology, 2200 N. Campus Blvd., University of Michigan, Ann Arbor, Mich. 48105

Mississippi Marine Advisory Program, P. O. Drawer AG, Ocean Springs, Miss. 39564

New Hampshire Marine Advisory Service, Kingsbury Hall, University of New Hampshire, Durham, N.H. 03824

New Jersey Marine Extension, Marine Science Center, Rutgers University, New Brunswick, N.J. 08903

New York Sea Grant Advisory Service, Fernow Hall, Cornell University, Ithaca, N.Y. 14850

North Carolina Marine Advisory Services, 134-1911 Bldg., North Carolina State University, Raleigh, N.C. 27606

Oregon Marine Advisory Program, 240 Extension Hall, Oregon State University, Corvallis, Oreg. 97331

Rhode Island Marine Advisory Service, University of Rhode Island, Narragansett Bay Campus, Narragansett, R.I. 02882

South Carolina Marine Extension Program, Marine Resources Center, P. O. Box 12559, Charleston, S.C. 29412

Texas Center for Marine Resources, Texas A & M University, College Station, Texas 77843

United States Department of Commerce, National Oceanic and Atmospheric Administration, National Marine Fisheries Service:

> Scientific Publications Staff, Room 450, 1107 N.E. 45th St., Seattle, Wash. 98105

> Northwest Fisheries Center, 2725 Montlake Blvd. East, Seattle, Wash. 98102

> Southeast Fisheries Center, 75 Virginia Beach Drive, Miami, Fla. 33149

> Northeast Fisheries Center, Woods Hole, Mass. 02543

> Southwest Fisheries Center, 8604 La Jolla Shores Dr., La Jolla, Calif. 92037

> Economic Research Division, 3300 Whitehaven St., N.W., Washington, D.C. 20235

> Extension Division, 3300 Whitehaven St., N.W., Washington, D.C. 20235

> Statistics and Market News Division, 3300 Whitehaven St., N.W., Washington, D.C. 20235

> Northwest Regional Office, 1700 Westlake Ave., North, Seattle, Wash. 98109

> Northeast Regional Office, Federal Building, 14 Elm St., Gloucester, Mass. 01930

> Southeast Regional Office, 9450 Gandy Blvd., North, St. Petersburg, Fla. 33702

> Southwest Regional Office, 300 South Ferry St., Terminal Island, Calif. 90731

> Alaska Regional Office, P. O. Box 1668, Juneau, Alaska 99801

Virginia Advisory Services, Institute of Marine Science, Gloucester Point, Va. 23062

Virginia Cooperative Extension Service, Virginia Polytechnic Institute and State University, Blacksburg, Va. 24061

Washington Marine Advisory Program, Division of Marine Resources, University of Washington, 3716 Brooklyn Ave. N.E., Seattle, Wash. 98195

Wisconsin Marine Advisory Services, University of Wisconsin, 420 Lowell Hall, 610 Langdon St., Madison, Wis. 53706

INDEX